Cities for a Shared Planet

One Planet

Sikina Jinnah and Simon Nicholson, series editors

Peter Dauvergne, *AI in the Wild: Sustainability in the Age of Artificial Intelligence*

Vincent Ialenti, *Deep Time Reckoning: How Future Thinking Can Help Earth Now*

Maria Ivanova, *The Untold Story of the World's Leading Environmental Institution: UNEP at Fifty*

Peter Friederici, *Beyond Climate Breakdown: Envisioning New Stories of Radical Hope*

Jennifer Clapp, *Titans of Industrial Agriculture: How a Few Giant Corporations Came to Dominate the Farm Sector and Why It Matters*

Thomas Princen, *Fire and Flood: Extreme Events and Social Change Past, Present, Future*

Gustav Peebles and Benjamin Luzzatto, *The First and Last Bank: Climate Change, Currency, and a New Carbon Commons*

Marina Alberti, *Cities for a Shared Planet: The Emergence of Multispecies Cooperation*

Cities for a Shared Planet

The Emergence of Multispecies Cooperation

Marina Alberti

The MIT Press
Cambridge, Massachusetts
London, England

The MIT Press
Massachusetts Institute of Technology
77 Massachusetts Avenue
Cambridge, MA 02139
mitpress.mit.edu

The MIT Press would like to thank the anonymous peer reviewers who provided comments on drafts of this book. The generous work of academic experts is essential for establishing the authority and quality of our publications. We acknowledge with gratitude the contributions of these otherwise uncredited readers.

This book was set in Stone Serif and Stone Sans by Westchester Publishing Services. Printed and bound in the United States of America.

Library of Congress Cataloging-in-Publication Data is available.

ISBN: 978-0-262-05555-0

EU Authorised Representative: Easy Access System Europe, Mustamäe tee 50, 10621 Tallinn, Estonia | Email: gpsr.requests@easproject.com

To Matteo

Contents

Series Foreword

This is at once an odd and exhilarating time to be alive. Our species, *Homo sapiens*, has had roughly 350,000 years on the planet. For most of that time, our ancestors barely registered as a quiet voice in a teeming chorus. No more. Now, a human cacophony threatens the ecological foundations upon which all life rests, even as technological wonders point the way toward accelerating expansion. We find ourselves at a moment of reckoning. The next handful of decades will determine whether humanity has the capacity, will, and wisdom to manufacture forms of collective life compatible with long-term ecological realities or whether, instead, there is an expiration date on the grand human experiment.

The One Planet book series has been created to showcase insightful, hope-fueled accounts of the planetary condition and the social and political features upon which that condition now depends. Most environmental books are shackled by a pessimistic reading of the present moment or by academic conventions that stifle a writer's voice. We have asked One Planet authors to produce a different kind of scholarship. This series is designed to give established and emerging authors a chance to put their best, most astute ideas on display. These are works crafted to show a new path through the complex and overwhelming subject matters that characterize life on our New Earth.

The books in this series are not formulaic. Nor are they Pollyannaish. The hope we have asked for from our authors comes not from overly optimistic accounts of ways forward, but rather from hard-headed and clear-eyed accounts of the actions we need to take in the face of sometimes overwhelming odds. One Planet books are unified by deep scholarly engagement brought to life through vivid writing by authors freed to write from the heart.

Thanks to our friends at the MIT Press, especially to Beth Clevenger, for guiding the One Planet series into existence and to the contributing authors for their extraordinary work. The authors, the Press, and we, the series editors, invite engagement. The best books do more than convey interesting ideas: They spark interesting conversations. Please write to us to let us know how you are using One Planet books or to tell us about the kinds of themes you would like to see the series address.

Finally, our thanks to you for picking up and diving into this book. We hope that you find it a useful addition to your own thinking about life on our One Planet.

Sikina Jinnah and Simon Nicholson

Preface

On my flight from Shanghai to Rome, during the second leg of a trip that circled the globe, I found myself immersed in a fascinating paper. The 2016 *Science* paper by Claire Spottiswoode presented experimental evidence of an extraordinary mutualistic relationship between honey hunters in sub-Saharan Africa and a wild bird species, the greater honeyguide. My son, Matteo, nine years old at the time, leaned in, captivated by the illustrations of the greater honeyguides and Mozambican honey hunters.

"What's that about?" he asked.

"It tells of a unique collaboration between humans and the greater honeyguide," I replied. I explained how the honey hunters in Mozambique use a special vocal call (*brrrr-hm* sound) to signal to the birds that they are seeking bees' nests, and the birds lead them to the hives by chattering using distinct, sharp sounds and flying from tree to tree. The hunters break open the hive with axes and subdue the bees with smoke, things the birds couldn't do on their own. Once the hunters have retrieved the honey, the birds eat the leftover wax and larvae.

"This partnership is rather special," I said. "Such deep interactions between humans and wildlife are rarely documented."

After a brief pause, Matteo asked, "Is it possible that humans and other species could form new collaborations in the future, Mom?"

Matteo's question lingered after he returned to his own reading, merging with my own thoughts about the keynote I had just delivered in Shanghai for the Second World Conference of the Society of Urban Ecology. My book, *Cities That Think Like Planets*, had just been released. In the book, I describe the growing evidence that cities are driving rapid evolutionary change in many populations of plants and animals. I show examples of how urbanization—by altering natural selection, neutral genetic drift, and gene flow—leads to shifts in species traits, further influencing ecosystem function, potentially changing evolutionary trajectories. With this planetary mindset, a question crossed my mind: Could the convergence of rapid urbanization and eco-evolutionary dynamics lead to novel forms of human–wildlife cooperation?

A vision started to crystallize.

I imagined a world in which humans cooperate with other species—birds, insects, fish, plants, fungi, and bacteria—to expand their ecological niche and their ability to survive and thrive in a rapidly changing environment.

What would such a world look like? Would people live in cities? Would there be futuristic metropolises? Would we live in villages? How would these urban systems function? What kind of technologies would enable multispecies cooperation? Is this a plausible future for human life on Earth—or perhaps the *only* one that would enable our species' survival?

Humans are able to imagine and realize the most ambitious explorations of the universe: finding new planets, determining whether they are habitable, and searching for other civilizations. But when it comes to our own planet, we tend to be much less ambitious. We have little confidence in the human capacity to evolve through symbiosis—a mutually beneficial relationship—with other organisms to create a new, more complex life on our own planet. In envisioning the future of Earth, our imagination tends to fall into a few simplistic archetypes based on skewed—and often extreme—simplifications of how nature works, anchored in biological determinism or technoscientific optimism. How can we overcome such a narrow view of our own ingenuity? Can we imagine a place where humans cooperate with other organisms to support each other's well-being? What kind of cities would we build in such a world?

Many books have eloquently described how cities are transforming the environment. Some recount the dramatic rise and fall of ancient urban centers and the downfall of entire civilizations. Others highlight the success stories of cities and their resilience over centuries. Several tackle difficult issues and offer solutions to safeguard the planet, its ecosystems, and its biodiversity from the impacts of urbanization.

This book offers a novel perspective. Rather than framing humanity and urbanization as primarily destructive forces, it views cities as active evolutionary systems in which humans and other species can evolve cooperative relationships. Drawing on emerging observations that cities act as drivers of eco-evolutionary change, I propose that urbanization represents a new stage in Earth's evolution. By exploring the convergence of accelerating urbanization and rapid evolutionary change, this book envisions the potential for a fundamental reconfiguration of human–nature relations, redefining the role humans play in shaping the future of our planet.

As our plane began to descend, I turned to Matteo. "Rome is the place that sparked my passion for cities."

He looked at me, curious.

"Among Rome's many monuments, the Roman aqueducts particularly captivated me when I was your age." I could still picture myself as a child, standing in awe before these immense structures, with my father explaining to my sisters and me that the Aqua Appia, built more than 2,000 years ago, was the first of a vast aqueduct system spanning 315 miles and supplying water to nearly a million people.

As a child, I was fascinated by the aqueducts' scale and the ingenious solutions employed by Roman engineers, such as the use of arches to exploit gravity. It was only later, as an urban ecologist, that I understood the broader significance of my experience. In the aqueduct, I saw humans as agents of niche construction, acting as ecosystem engineers, shaping their environment for survival, much like other species. This insight led me to see the potential for bridging the gap between biology and urban design. The perspective of humans as planetary engineers has inspired my work and studies ever since.

My journey around the globe with Matteo back in 2016, where the seeds of this book began to germinate, prompted a shift in my thinking. Beyond the ways humans adapt their environment, I began to consider the potential for humans and other species to cooperate. Stops on that journey—Venice, with its advanced salt-marsh restoration and experimental flood-control strategies; Berlin, with its pioneering green rooftops; and London, in the process of restoring its urban wetlands—began to shine in a new light. I noticed signs of an emerging synergy between humans and nature. Each innovative green and blue infrastructure I encountered signaled progress toward a more sustainable future and hinted at the prospect of humans living in symbiosis with other organisms. I began to envision principles of multispecies design and bio-cooperation.

Although the prevailing view in ecology still sees cities as inherently opposed to nature and as the cause of our environmental problems, emerging empirical evidence and changes in perspective, particularly among younger scholars, are starting to alter this narrative. Urbanization doesn't necessarily signify the end of nature. Instead, it can represent a transformative phase in the planet's evolutionary trajectory, redefining our relationship with the environment.

The story of the honeyguide inspired this book and prompted me to ask questions we had not yet asked about Earth's future. Evolution—an open-ended process—can give rise to new and diverse forms of life over time. To see cities through the lens of evolution requires a fundamental shift in our perception and understanding of our place in nature.

As I conclude *Cities for a Shared Planet*, I reflect on my recent visits to Singapore—a city that continues to inspire my imagination and exemplify the potential for a new partnership between humans and other species. On my latest visit, I observed not only tangible progress but also the ambitious vision driving its evolution into a "City in Nature," where nature is intentionally integrated into the urban fabric. Riding along the Kallang Park Connector, part of Singapore's 300-kilometer trail network, observing a once-paved river now restored to its natural state in Bishan-Ang Mo Kio Park, and walking through downtown's vertical gardens, I saw nature reemerge—not by chance, but through deliberate design as an experiment in coevolution that marks a new chapter for the future of cities. This vision of multispecies urban coexistence redefines urban design as the co-creation of shared habitats where humans and other species live in symbiosis. Singapore offers a living example of what cities can achieve when nature is embraced as an essential partner in shaping resilient and inclusive futures.

A sustainable hybrid planet *is* possible if we have the imagination to rethink human–natural relationships as true partnerships.

Seattle, June 13, 2025

Marina Alberti

Acknowledgments

This book is inspired by my son's curiosity and imagination. Through his eyes, everything becomes possible.

My students have been central to this project. In classrooms, studios, and fieldwork—and through sustained, thoughtful conversations—they challenged assumptions, surfaced hidden dimensions, and gave form to ideas through real-world application.

I am especially grateful to the Urban Ecology Research Lab. Successive cohorts have nurtured an intellectual environment rooted in curiosity, rigor, and collaboration.

A heartfelt thanks goes to the students in my first experimental multispecies design lab course for their thoughtful engagement with the ideas developed here: Keisha Bharath, Chin-Wei Chen, Matt Grosser, Aaron Julius Morado Lecciones, Ziyang Liu, Akeo Maifeld-Carucci, Anna Malesis-Dahm, Amy Lisa Miller, Siman Ning, Hunter Ottman, Natalia Perdomo Ayerdi, Hennessy Elena Rodriguez, Laura Alexis St. Jarre, Chelle Tobey, and Brendan Wallace.

Many people have contributed to this book in crucial ways—far more than I can name here. I am especially grateful to the mentors and long-standing collaborators across ecology, evolution, planning, design, and the social sciences who have shaped the questions I ask and the standards to which I hold myself.

This book is grounded in evolutionary thinking that treats organisms as active participants in shaping life's conditions. I am indebted to Lynn Margulis for placing symbiosis and cooperation at the heart of evolutionary change; to Richard Lewontin for emphasizing feedback, contingency, and the inseparability of organisms and environments; and to Kevin Laland for formalizing niche construction as an evolutionary process. I am also grateful to scholars in eco-evolutionary dynamics whose work has transformed how contemporary evolutionary change is studied and interpreted, including Andrew Hendry, David Reznick, and Thomas Schoener. Their collective work provides the intellectual bedrock on which this book stands.

The NSF Research Coordination Network on urban eco-evolutionary dynamics (Award No. DEB-1840663), which I had the privilege to lead, further deepened my engagement with recent advances in evolutionary science. I am grateful to the many collaborators in this network whose insights sharpened how I think about feedbacks, genetic and phenotypic trait variation, and evolutionary change in urban systems.

The mutualistic relationship between honey hunters in sub-Saharan Africa and honeyguides, documented by Claire Spottiswoode and her collaborators, served as a formative inspiration for this book, prompting questions I had not previously considered. I'm grateful to Claire for her generosity in discussing her research and its broader implications, as well as to Dominic Cram and Jessica van der Wal for conversations that shaped how this case is developed here.

This book draws on a body of multispecies scholarship that has expanded how agency, interaction, and meaning are understood beyond the human. The work of Donna Haraway, Anna Tsing, and Bruno Latour—alongside contributions in geography and urban planning by Sarah Whatmore and Donna Houston—provided essential points of reference.

My thinking on multispecies design has been enriched by stimulating conversations with scholars across disciplines, including Danielle Celermajer, Christoph Rupprecht, and Wolfgang Weisser. I am especially grateful to my colleague and friend John Marzluff, who grounded these explorations in decades of empirical work on urban wildlife.

I also thank Jason Munshi-South, Carly Ziter, Kristien Brans, and Chris Schell, whose research and field experience in urban ecology and urban evolution, and whose reflections—shared with me through interviews—on the excitement and complexity of studying evolution in cities, have informed and strengthened the perspectives shared in this book.

I am grateful to Puay Yok Tan for generously sharing his perspective on the evolution of Singapore's environmental strategy and for clarifying how the shift from *City in a Garden* to *City in Nature* reflects deeper commitments to biodiversity, ecological connectivity, and long-term socioecological monitoring. I also thank Perrine Hamel for conversations that placed Singapore's experience in a broader comparative context, highlighting both its experimental strengths and the limits of direct replication elsewhere.

I thank Beth Clevenger, editor at the MIT Press, for believing in this project and for guiding it with clarity and care, as well as the reviewers whose encouragement and thoughtful critiques strengthened both the argument and its articulation. I am also grateful to editor Susan Matheson and to Micaiah Evans, whose editorial interventions at key moments sharpened the language while preserving the complexity of the ideas. Tianzhe Wang provided additional editing, verified reference accuracy, and formatted the manuscript with extraordinary dedication and care.

Thanks also to the production team for their thoughtful attention in bringing this book to life and to graphic designers Sandy Westermann and Madelyn Neufeld, who skillfully translated key concepts into visual form.

I close with gratitude and a sense of wonder for this extraordinary planet we inhabit and the vast cosmos it drifts within. Nature's capacity to surprise, to transcend our understanding, and to invite ever-deeper inquiry is a source of enduring inspiration and hope. The future of humanity will be defined by how we meet nature's generosity, with equal humility, care, and imagination.

Introduction: The Dawn of Bio-Cooperation

Theo Yoon scrolled through the archived footage, the screen before him alive with the flickering images of a city once on the brink of collapse. His office overlooked the city, a view that had been drastically transformed over the past century. The deadline for his piece loomed, and he felt the weight of the challenge ahead.

He had been assigned to write a story for the *New York Times*—a reflection on the 100th anniversary of 2050, the year that marked a turning point in the battle against climate change. This article would transcend mere historical recounting; it was a chance to capture the essence of a century-long metamorphosis that had reshaped New York City. A hundred years ago, the city was on the edge. Now, in 2150, it is a thriving metropolis, a seamless blend of urban and natural life.

Neighborhoods function as micro-ecosystems, where urban parks and nature corridors create thriving habitats for wildlife while simultaneously offering peaceful refuges for urban dwellers. Community gardens have become hubs of social connection and sanctuaries for pollinators. Innovative bioswales and rain gardens, seamlessly integrated into the urban landscape, manage stormwater while fostering diverse ecosystems of plants, insects, and amphibians. Birds share the skies with drones, contributing to a sophisticated network for transportation, communication, and environmental monitoring, exemplifying the city's harmonious blend of technology and nature.

The Hudson River has been naturalized. Coastal wetlands now line its shores, serving as filters for contaminants and providing a home for water-dwelling organisms. Beneath the water's surface, extensive oyster reefs—accreted over decades—now stabilize shorelines, filter the harbor's waters, and structure complex food webs, functioning as a living substrate that supports the long-term ecological recovery of the estuary. Along the riverbank, cutting-edge floating residential areas have emerged, engineered to adapt to tidal fluctuations and to withstand increasing ocean levels. These water-based settlements enhance the city's capacity to cope with climate change, seamlessly integrating human innovation with ecological processes. Along these waterways, river otters serve as sentinels of ecosystem health, reflecting water quality and ecosystem functioning.

Jamaica Bay marshlands provide a sanctuary for migratory birds, fish, and countless other species. The city has embraced its role as a steward of this vital ecosystem, recognizing that its health is intertwined with the well-being of all New Yorkers. Embedded in the wetland soils, microbial fuel cells harness the metabolic processes of microbes to generate electricity, contributing to the city's power supply while enhancing the ecological function of these areas. In Jamaica Bay, the urban and the wild coexist in a delicate balance, each supporting the other in a complex web of life.

The city's architecture has become an extension of its natural surroundings. Buildings are adorned with vertical gardens that offer food and shelter to birds and insects, while rooftop wetlands purify rainwater and support amphibian life-cycles. Beneath the streets, a network of mycorrhizal fungi connects tree roots, enhancing nutrient distribution and soil health. Aquatic corridors weave through the urban fabric, supporting fish populations integral to natural stormwater management. In public spaces and on building facades, algae bio-reactors sequester carbon dioxide, while microbial fuel cells embedded in the infrastructure generate clean electricity, making these structures key contributors to the city's sustainable energy grid.

Plants equipped with bio-sensing capabilities monitor the urban microclimate, providing real-time data on air quality and weather patterns that inform both urban planning and daily life. Every structure, every space within the city, is designed to foster and enhance these interspecies relationships, making New York a testament to the possibility of human–nature coexistence.

On Yoon's screen, the title of his piece came into focus: "Climate Change: The Wave That Turned the Tide."

The turning point came in 2050, a year that would later be recognized as the moment when humanity made a crucial choice. The convergence of rapid urbanization and intensifying environmental pressures made it clear that adapting to the immediate challenges of climate change was no longer enough. A fundamental shift in perspective was essential to develop a novel strategy for enhancing urban resilience. The transformation did not happen overnight. It was the result of years of challenges, failed interventions, and crises. As the city struggled with ongoing threats, a turning point emerged, driven by a simultaneous shift in values. The realization had dawned that overcoming the climate crisis required more than technological fixes or isolated green initiatives. It demanded a fundamental rethinking of cities as not just human habitats but shared spaces where the needs and dynamics of other species were integrated into the urban fabric.

This new perspective—one that strategically aligned human development with the ecological needs of other species—set New York on a path to becoming the resilient, thriving metropolis it had become. It marked the dawn of a new symbiotic relationship between humans and nature.

Can this imagined transformation that turned New York City into a bio-cooperative city emerge in today's world?

Imagination can push the limits of what we consider possible. Yet, even when set in the future, the archetypes in science fiction and other speculative genres often reflect and oversimplify present-day realities. Utopian and dystopian narratives frequently assume a fundamental divide between humans and nature—where humans either dominate nature or are overwhelmed by it. This narrative has so deeply shaped our identity that even our visions of future worlds are biased toward the same human dilemmas.

This book reveals another story. I propose that the convergence of rapid urbanization and accelerating environmental change may signal the onset of a major evolutionary transition on Earth: the emergence of a novel symbiotic relationship between humans and nonhuman organisms as an eco-evolutionary strategy for resilience. Central to this transition is the rise of bio-cooperative cities: adaptive, multispecies systems in which human development is informed by the ecological needs and evolutionary potential of other organisms, setting the stage for a new era of ecological urbanism. This vision is grounded in bio-cooperation, cooperative interactions between human and nonhuman species shaped by eco-evolutionary feedbacks. This perspective shows how integrating the ecological needs and agency of other species into urban systems can promote more resilient and thriving cities.

Drawing on urban ecology, evolutionary biology, design theory, and environmental justice, the book integrates empirical evidence with conceptual innovation to reframe cities as coevolving ecosystems—places where humans and other species influence each other's trajectories through ongoing ecological and evolutionary interactions. It advances the idea that resilience amid climate disruption and biodiversity loss depends on embracing ecological interdependence as a core principle of design. Cities become shared ecological niches, continually co-constructed through processes of mutual adaptation and cooperation.

Through real-world case studies and imaginative scenarios, the book outlines key properties that define bio-cooperative cities, providing a blueprint for reenvisioning urban design.

Observations from cities such as Seattle, Singapore, New York City, Ho Chi Minh City, and Cape Town demonstrate how multispecies interactions are transforming infrastructure, reshaping urban design, and redefining how urban environments are inhabited and evolve.

Global urbanization is the most visible signature of the Anthropocene, the current geologic epoch, which defines our species as a major influence on the planet.[1] While humans have been altering ecological processes for millennia, urbanization represents an unprecedented shift in intensity, speed, and scale.[2] Our technological capacity has vastly expanded the human ecological niche, both in its impact and its dependence on the environment, positioning our species as planetary engineers. Through our cities, we have not only redefined ecosystems but also altered their relationships with both nearby and distant environments. In these urban landscapes, the tightly coupled interactions between humans and other organisms are challenging species' survival, reproduction, and fitness, altering traits and leading to the emergence of new species.

In the past decade, evolutionary biologists have documented compelling evidence of urban-driven evolutionary changes across a wide array of species, including birds, fish, plants, mammals, and invertebrates. Songbirds fine-tune their calls to navigate the cacophony of urban noise, flowering weeds like *Crepis sancta* modify their seed dispersal to survive on impermeable surfaces, and water fleas (*Daphnia*) alter their physiology to withstand chemically altered aquatic environments.[3] Urbanization isn't just driving evolutionary change in individual traits within populations; it's restructuring the evolution of pace-of-life syndromes—fundamental shifts in survival and reproductive strategies.[4] These changes have the potential to reshape ecosystems, with significant consequences for human well-being on contemporary time scales.[5]

Cities have become the new ecological stage on which the evolutionary play is being performed.[6] What Charles Darwin observed in the Galápagos Islands—the process of evolution by natural selection—is now clearly visible much closer to home. In 1835, when Darwin visited the Galápagos, the planet had just over 1.1 billion people, with only about 5.4% living in urban areas.[7] Today, over 4.5 billion (56%) of the planet's 8 billion people live in cities, a proportion expected to reach nearly 70% by 2050.[8]

Urbanization is driving systemic changes in social and ecological systems, accelerating the rates of interaction among people and places and strengthening the connections between human activities and Earth's ecosystems, expanding the influence of human agency on fundamental processes that sustain life.[9] As humans reshape the landscape, urbanization accelerates evolutionary change in species vital to ecosystem functions, tightens eco-evolutionary coupling, and alters resilience, influencing both local and global sustainability over the long term.[10] These eco-evolutionary feedbacks, once only hypothesized, are now being observed empirically.[11] Evolutionary biologists are beginning to uncover the potential consequences of the reciprocal and simultaneous interactions between environmental change and rapid evolutionary shifts in species traits.[12] Understanding these dynamics is crucial for predicting trait changes in urban environments and for facilitating urban adaptation to climate change.

Throughout this book, I use the term adaptation in two distinct ways. In biological contexts, it refers to heritable evolutionary change in populations driven by natural selection. In urban and societal contexts, it refers to planned social, institutional, and technological adjustments that reduce climate-related risk and vulnerability.

Cities offer a unique opportunity to observe evolution in real time through global, replicated experiments.[13] Despite their differences, cities allow us to ask whether populations exposed to similar selection pressures undergo similar evolutionary changes and how multiple selection pressures might affect organisms' ability to adapt.[14] Global variation in city form and scale allows us to disentangle which urban properties shape evolutionary outcomes.

Human-driven eco-evolutionary feedbacks pose new challenges to Earth's ecosystems and simultaneously create opportunities for evolutionary innovation. These feedbacks could reshape the interactions between our technologically driven civilization and global ecological processes, leading to unprecedented biospheric functions, such as novel free energy-generation capacities.[15] As we become aware of our power to influence evolution, these emerging eco-evolutionary feedbacks might prompt new cooperative relationships between humans and other species, potentially leading to the evolution of a *hybrid* planet.[16] In this book, I explore two central questions: (1) Does the global rise of cities signify a major evolutionary transition comparable to the Great Oxidation Event, when a sharp increase in atmospheric oxygen enabled the evolution of complex life on Earth?[17] (2) Can global urbanization foster new symbiotic relationships between humans and other species?

To address these questions, I propose four hypotheses that provide a framework for systematic exploration. First, I suggest that urban environments might exhibit distinctive eco-evolutionary feedbacks detectable across multiple species and ecosystem functions. Cities often show a clear signal of phenotypic change—observable changes in organismal traits—across many species and regions.[18] These urban-driven changes can affect population dynamics and, in turn, reshape community dynamics and ecosystem functions. I explore how the co-occurrence of multiple selection pressures in cities might influence species' responses to individual stressors or generate synergistic interactions that amplify or dampen eco-evolutionary feedbacks.[19]

The second hypothesis is anchored in the potential emergence of eco-evolutionary feedbacks. If the phenotypic changes observed in urban regions are genetically based, they could give rise to eco-evolutionary feedbacks with consequences for ecosystem function and stability. For instance, urbanization might influence the evolution of traits in primary producers and decomposers—such as growth rates, nutrient uptake efficiency, and metabolic traits—shaping key ecosystem functions, including primary productivity and nutrient cycling, with profound implications for human well-being. These insights are essential for informing urban planning and resilience strategies.[20]

Third, I hypothesize that the convergence of global urbanization and accelerating environmental change can catalyze rapid evolutionary dynamics, fostering novel, complex symbiotic relationships between humans and other organisms, potentially reshaping Earth's evolutionary trajectory.[21] The accelerated pace and distinctive challenges of urban life, coupled with a new societal awareness driven by rapid environmental change, could create conditions for novel forms of cooperation between humans and other species to emerge. In this scenario, urban pressures might accelerate the evolution of traits that reinforce functional interdependence among species and support ecosystem resilience. Though few examples of such cooperation currently exist,[22] they may provide a glimpse into the potential relationships that could emerge within cities.

Finally, I hypothesize that by supporting a wide variety of life forms, *multispecies design* can enable more biodiverse and resilient human habitats. Traditional urban planning has often prioritized biodiversity that directly serves human needs, overlooking the requirements of other urban inhabitants such as birds, insects, and plants. In contrast, a multispecies perspective considers the well-being and agency of all urban species, supporting biodiversity conservation and coexistence. This approach suggests that the deliberate inclusion of multiple species in urban design could enhance the quality of life for both humans and nonhumans.

Drawing on real-world examples and hypothetical scenarios grounded in current scientific understanding, I outline the foundations of what I describe as bio-cooperative cities. These cities exemplify a new, cooperative relationship between humans and other organisms, aimed at sustaining ecosystem function and well-being in the long term.

To create resilient cities, we must recognize our crucial role in Earth's evolution. As active agents of niche construction, humans profoundly alter the environment and the selection pressures that shape our own evolution and that of other species. In cities, we change everything—composition, processes, dynamics, and functions.[23] By constructing skyscrapers, stormwater drains, road networks, and subways, we redistribute organisms and alter energy and material cycles, leading to unique urban biogeochemistry, biodiversity, and ecological functions. Like other species, humans engage in biotic interactions that underpin the structure and function of ecological communities. Recognizing this fundamental similarity demands an understanding of niche construction and cultural coevolution, shaped by both inheritance and social learning.[24]

As we face some of the most profound challenges in human history, we confront the ethical and practical imperatives of our role on this planet. We are not passive observers of Earth's evolution but active participants shaping its trajectory. How can science help us align our actions with these responsibilities? Moving from theory to actionable science is no longer optional—it is essential. Urbanization, humanity's greatest unintended evolutionary experiment, reveals how life might adapt to future environmental change. Modern cities, microcosms of a fully urbanized planet, provide an unparalleled platform for comparative scientific inquiry. They illuminate the spectrum of human decision-making and its influence on planetary trajectories, enabling us to decode eco-evolutionary dynamics at an unprecedented scale. By integrating diverse human perspectives with the adaptive strategies of other species, we can chart a path toward resilient, sustainable urban environments. The generation of knowledge is no longer the domain of science alone—it is a collective endeavor that integrates human creativity with nature's ingenuity, guiding us toward a sustainable future.

I propose that bio-cooperative cities arise from the interaction between rapid social change and rapid evolution. These cities balance the need to preserve important aspects of the environment with the necessity to innovate and adapt. In living systems, resilience requires both robustness and flexibility. From an evolutionary perspective, robustness is essential for an organism to maintain the stability of traits required for fitness, whereas flexibility allows for innovation, enabling the organism to adapt to new environmental challenges. The interplay between conservation and innovation has been shown to enable systems to process information quickly, respond effectively to disruptions, and adapt to changes without losing their essential functions.[25]

The vision of a bio-cooperative city goes beyond merely redesigning a city's physical structure and functions. It requires a shift in our mindset to conceive and design cities as habitats for multiple species. This demands the integration of human needs with those of other species, reimagining the city as shared habitat. It calls for an understanding of our intrinsic role in nature's interconnected web of life and the harnessing of our ability to influence planetary evolution, steering our cities toward a future where all species coexist and thrive. Communities become stewards not only of nature but also of the symbiotic relationships between humans and nature. Society must invest in research to unravel the intricate relationships between humans and other species, as well as empower communities to contribute to knowledge production and strategy development for achieving desirable outcomes. Embracing uncertainty and accepting the limits of our knowledge, governments should take proactive measures, promote environmental education, and prioritize policies that strengthen multispecies relationships.

The transformative shift in human consciousness described here may seem unlikely, even unimaginable, unless considered through an evolutionary lens. Emerging insights from population genetics on niche construction suggest that this process can drive evolutionary innovation and stability, leading to uncommon dynamics and coevolution. I argue that our current epoch presents a unique opportunity for systemic change, prompted by the convergence of major transitions in both human society and Earth's systems. This convergence could trigger a radical transformation in our relationship with nature.

In this work, I explore the dynamic interplay between urban ecosystems and evolutionary processes, considering how interspecific mutualism may shape hybrid urban systems. This perspective requires clear boundaries.

While proposing that the emergence of eco-evolutionary dynamics has the potential to transform relationships between humans and other species, we must be cautious not to overextend the evolutionary paradigm. Evolutionary dynamics provide insights into the emergence of novel phenotypes and coevolutionary processes, but they do not, on their own, resolve all social or environmental challenges. Other scientific disciplines, social theories, and cultural considerations play vital roles in forming a comprehensive understanding.

Applying an eco-evolutionary lens is not meant to offer an all-encompassing explanation for human–environment interactions or social phenomena. Rather, it provides a perspective that refines our understanding of the complex relationships between humans, other organisms, and the urban environment. It highlights the innovative potential of eco-evolutionary dynamics in shaping new ecological pathways.

The hypothesis of bio-cooperative cities presented in this book, situates evolutionary processes within a broader ecological and social context. By integrating evolutionary insights with other scientific disciplines, social theories, and cultural perspectives, it creates common ground for collaboration essential to understanding the complex human–environment relationships of the Anthropocene.

This book is organized into two parts:

Chapters 1 through 4 lay the conceptual and scientific groundwork for understanding the dynamics that shape bio-cooperative cities, revealing the eco-evolutionary processes that guide their transformation. These chapters examine four core dynamics—evolution, complexity, symbiosis, and resilience—that drive how urban environments function, adapt, and thrive as shared habitats for human and nonhuman life. These dynamics establish the foundations from which the emergent properties and guiding principles of bio-cooperative cities arise.

Chapter 1 examines how urban environments drive rapid evolutionary change, highlighting shifts in species traits and behaviors that reshape survival strategies, fuel adaptation, and create new ecological interactions. Chapter 2 explores the complexity of urban ecosystems, showing how urbanization rewires ecological networks, creates novel niches, and alters evolutionary trajectories in ways that challenge conventional views of species' ecological roles. Chapter 3 focuses on the potential for symbiotic relationships in cities, proposing that urban environments can foster new forms of cooperation between humans and other species. Chapter 4 redefines urban resilience through an eco-evolutionary lens, emphasizing the need for dynamic urban planning that supports evolutionary adaptation in response to environmental change.

Chapters 5 through 11 build on these foundations to explore the emergent properties of bio-cooperative cities—distinct qualities that arise from dynamic interactions between human and nonhuman organisms in response to environmental change. They examine how these dynamics reshape cities into shared ecosystems that support both human and nonhuman life. Chapter 5 introduces five key properties—evolving urban niches, eco-evolutionary dynamics, multispecies equity, the interdependent properties of evolvability, adaptability, and foresight, and cooperative innovation—setting the stage for deeper exploration in the following chapters.

Chapter 6 investigates how cities function as evolving, co-constructed niches, where human and nonhuman interactions shape novel habitats that promote coexistence. Chapter 7 examines the role of eco-evolutionary dynamics in urban planning, emphasizing feedback systems that enable cities to adapt to environmental change. Chapter 8 advances a framework for multispecies equity, recognizing the agency of nonhuman species and their integral role in urban systems. Chapter 9 distinguishes the interdependent properties of evolvability, adaptability, and foresight, showing how cities can leverage biological anticipatory capacities for long-term resilience.

Chapter 10 highlights how cooperation between humans and wildlife can drive innovation, proposing new forms of mutualism as plausible evolutionary transitions. These threads converge in Chapter 11, where I synthesize the principles of bio-cooperative design into a cohesive framework, that envisions urban environments as adaptive, cooperative ecosystems.

This book is an invitation to future city planners to embrace a vision of cities and nature thriving in symbiosis. By imagining cities where humans cooperate with other species, we can uncover new principles to create what I call bio-cooperative cities. Applying multispecies design principles and harnessing the eco-evolutionary potential of nature-based solutions, cities can develop innovative strategies to enhance resilience and address the challenges of climate change.

1 Evolution in Action

Imagine acquiring the sensory capacity to experience the world as one of the myriad organisms that share our urban spaces. Last spring, in my urban ecology class, I challenged my students to see the city from the point of view of a nonhuman organism, to explore the invisible world inaccessible to humans through their senses, what Jakob von Uexküll calls the *umwelt*.[1] Working in teams, students were tasked to conceptualize an organism's *phenomenal world* and imagine experiencing the city based on the organism's unique sensory capacity, physical and biological makeup, behavior, and interactions with its surroundings (figure 1.1).

By imagining a lizard's tactile experience, one team described the struggle of climbing a metal pole or a glass-sided building. The metal pole is slick beneath the lizard's feet, and the glass-sided building is even slipperier. Every climb feels like an impossible challenge. Another team explored songbirds' auditory perception, imagining the overwhelming urban noise from a bird's perspective. The city's noise hinders the birds' attempts to communicate, attract mates, and defend territories. This auditory immersion highlighted how urban noise forces these birds to adapt their calls—altering pitches and volumes—to be heard.

Another team of students looked at how bees are forced to navigate the city's vertical gardens and can be baffled by flowers growing atop 900-foot skyscrapers. One team imagined being a weed; when a seed from the weed lands on a paved parking lot, it finds no soil to root in, leaving little hope for growth.

Through this exercise, it became clear that organisms in cities have no choice: adapt or go extinct. It is important to distinguish between adaptation and evolution. Adaptation can occur through phenotypic plasticity—nongenetic changes in behavior, physiology, or morphology that allow

Figure 1.1
Sensory worlds of urban species: different organisms perceive and navigate the city through distinct sensory modalities. Source: Author-created drawing. Photographs: Feral Pigeon (*Columba livia*), iNaturalist, Stephen James McWilliam, public domain (CC0). Western Honey Bee (*Apis mellifera*), iNaturalist, Joshua, public domain (CC0). Bat (*Myotis daubentonii*), iNaturalist, kdramsay29, public domain (CC0). Lichen (*Parmotrema reticulatum*), iNaturalist, Michelle Lopez, public domain (CC0). Ant (*Formica* spp.), iNaturalist, Andrew Cunnington, public domain (CC0). Crow (*Corvus* spp.), iNaturalist, Calinsdad, public domain (CC0). Spring Peeper (*Pseudacris crucifer*), iNaturalist, Sara L. Giles, public domain (CC0). Largemouth Bass (*Micropterus nigricans*), iNaturalist, Kathleen Houlahan Chayer, public domain (CC0). Daphnia (*Daphnia* spp.), iNaturalist, Victor Heng, licensed under CC BY 4.0. Oak tree (Quercus spp.), Flickr, Worthing Wanderer, public domain (CC0).

organisms to cope with new environments. Evolution involves heritable genetic change across generations. While both can produce similar functional outcomes, only evolutionary change alters the genetic composition of populations and contributes to long-term lineage divergence.

Since Charles Darwin first observed natural selection, evolutionary biologists have viewed the Galápagos Islands as a microcosm of natural evolution. The diverse and isolated environments of the Galápagos provided

Darwin clear insights into how species adapt and evolve in response to distinct environmental pressures. Today, cities represent a modern microcosm of evolution, which can be observed in real time. Cities pose unique challenges to a diverse array of organisms, driving rapid evolutionary changes in traits of many plants, animals, fungi, insects, and microbes. These adaptations influence ecosystem functions that support human well-being, such as pollination, nutrient cycling, climate mitigation, and carbon sequestration.

Eco-evolutionary feedbacks—the dynamic interplay between ecological and evolutionary processes—can occur over remarkably short time scales. Increasing evidence shows that evolutionary responses to strong urban selection pressures—such as habitat fragmentation, temperature shifts, or pollutants—can emerge within just a few generations. This challenges the long-standing assumption that evolution unfolds too slowly to matter for contemporary ecological dynamics, revealing instead a tightly coupled feedback between ecology and evolution observable on contemporary time scales.

A well-documented example of rapid evolution involves the water flea *Daphnia*. This common zooplankton, found in urban lakes and ponds, plays a pivotal role in freshwater food webs and has recently evolved to tolerate cyanobacteria, higher temperatures, and road salt.[2] Another example is the flowering weed *Crepis sancta*, similar to the dandelion, which disperses fewer seeds in urban areas than in the countryside. A study shows that urban weeds growing in Montpellier, France, disperse fewer of their seeds than they did 12 years prior.[3] Urban-dwelling house finches (*Carpodacus mexicanus*) have evolved larger, stronger beaks to feed on sunflower seeds from city bird feeders, seeds that are markedly different from the native seeds of the Sonoran Desert that they originally consumed.[4] Similarly, earthworms (*Lumbricus rubellus*) in urban settings have genetically adapted to various human-induced soil contaminants.[5] Atlantic killifish (*Fundulus heteroclitus*) in urban estuaries have adapted to tolerate elevated levels of toxic contaminants, such as polychlorinated biphenyls, heavy metals, and polycyclic aromatic hydrocarbons.[6] All these evolutionary changes (which are described in more detail below) can alter trophic interactions, population dynamics, and biogeochemical cycles, thereby affecting ecosystem functioning.[7]

Selection pressures in cities can be intense, changing organisms' survival rates and reproductive success. These pressures operate through several evolutionary mechanisms. Natural selection favors individuals with traits suited to urban conditions. Genetic drift can lead to random changes in gene frequencies, especially in small, fragmented populations. Gene flow—the movement of genes between populations—may be restricted or redirected by urban infrastructure, altering evolutionary trajectories. Mutation introduces new genetic variation into populations, providing the raw material on which natural selection can act.

The changes in gene frequencies that lead to trait changes in organisms living in cities are what I call *urban phenotypes*. These changes can cascade down levels of ecological organization—from populations to communities and ecosystems—and ultimately affect evolution through both genetic (changes in gene frequencies) and cultural coevolution (social learning). Organisms are adapting to the new environment by changing their physiology, morphology, life history, and behavior, reshaping ecological communities, species interactions, and overall ecosystem dynamics.

The Urban Phenotype

Scientists have documented many examples of rapid evolution associated with urbanization, highlighting their significance for human well-being. These adaptations have the potential to influence many ecosystem functions, such as pollination, food production, carbon sequestration, climate mitigation, pest control, and human health. Their impacts are occurring at rates that have the potential to affect our present lives. Understanding these evolutionary adaptations is essential for maintaining ecosystem function and safeguarding the health and resilience of urban communities.

The Urban Water Sentinel

The water flea *Daphnia* plays a significant role in food webs. Despite their microscopic size, these water fleas—found in urban lakes—have an enormous biological effect on freshwater ecosystems. As highly efficient grazers of phytoplankton, *Daphnia* help control lake eutrophication by eating algae and bacteria. Eutrophication happens when a lake accumulates an excess of nutrients, causing algae to grow out of control. Although toxic compounds produced by cyanobacteria can be harmful to zooplankton like *Daphnia*, exposure can select for more tolerant genotypes within *Daphnia* communities, leading to trait changes.

One of the most common species of *Daphnia* is *Daphnia pulex*. In response to increased algae blooms, *Daphnia pulex* has evolved to tolerate cyanotoxins produced by cyanobacteria, impacting feedback mechanisms that determine the net effect of eutrophication. The evolution of *Daphnia* shows how a tiny organism can adapt to nutrient enrichment in urban water bodies through urban wastewaters and stormwater runoff, with implications for the entire ecosystem. *Daphnia* has also been shown to rapidly adapt to urban heat and road salt, although interactions among multiple stressors may affect *Daphnia*'s adaptation. *Daphnia magna*, another common species, has evolved increased heat tolerance and changes in life history traits in urbanized habitats. *Daphnia galeata* has rapidly adapted to road salt, potentially buffering trophic cascades associated with lake salinization that might otherwise reduce grazing pressure and elevate phytoplankton levels.

Daphnia adaptation is one of the best-documented examples of human-driven eco-evolutionary change, making it an excellent system for studying urban eco-evolutionary dynamics due to its important role in freshwater food webs. *Daphnia*'s role in regulating phytoplankton has been central to efforts to control eutrophication through biomanipulation.[8] An eco-evolutionary perspective prompts the question of whether evolutionary responses in *Daphnia* populations—such as adaptation to cyanobacteria or urban stressors—can inform nature-based approaches to algal bloom management.[9] While evidence from large-scale interventions remains mixed and context dependent,[10] studying these adaptive responses can deepen understanding of eco-evolutionary dynamics relevant to environmental management.

The Sedentary Seed

Weeds that grow in small habitat patches around street trees and sidewalks produce seeds that have evolved to travel less far than their countryside counterparts. *Crepis sancta*, a daisy-like annual weed, has two types of seeds: heavy ones that drop straight down and lighter ones that can be caught on the wind. *Crepis sancta* growing in cities quickly evolved to produce more of the heavy seeds, which have a higher chance of falling on suitable ground for germination. A long-term study in Montpellier, France, confirmed this trend, documenting a significant reduction in seed dispersal among urban weeds over just 12 years.[11]

Environmental changes caused by urbanization affect plants' seed dispersal traits and their dispersal agents. Habitat fragmentation, increased temperature, nutrient depletion, and the loss and redirection of dispersal vectors in cities influence seed dispersal traits, often increasing the proportion of nondispersing seeds and altering dispersal distance and seed size. Studies have shown that seed dispersers exert directional selection on specific fruit traits, which in turn influence patterns of seed dispersal.

In cities, humans also act as agents of dispersal, altering spatial networks, patterns, and distances of dispersal.[12] Vehicles have significantly changed seed dispersal mechanisms in urban areas, facilitating the spread of nondispersing seeds, sometimes over long distances. New dispersal pathways transfer nonnative species from urban to rural habitats, leading to invasive plants in the suburbs. Scientists collected seed samples at highway tunnel entrances in Berlin to study the effect of road traffic on plant diversity. They found that road traffic facilitates the "suburbanization" of species predominant in urban habitats to suburban and rural areas, potentially leading to the homogenization of plant species across an urbanizing landscape.[13]

As humans become agents of seed dispersal, they reshape competition among species by altering dispersal pathways and survival opportunities. The structures and infrastructure they build create new vectors, corridors, and habitats that influence how far seeds travel and their establishment success. While fragmentation is perhaps a key selective agent for dispersal traits, it also affects plant reproductive traits and plant size. For many plants, reproduction depends on pollinators. Urban-driven changes in the abundance and diversity of pollinators may lead to floral trait evolution, selecting for increased floral attractiveness or longevity, as well as reproductive traits that rely less on pollination and more on self-fertilization.

The Kind Earthworm

Earthworms, commonly encountered on sidewalks during rainy days, play a crucial role in terrestrial ecosystems by recycling organic matter and processing new toxins. In urban backyards and parks, they are extra helpful. They process pollutants in the soil. They can tolerate extremely high concentrations of heavy metals, and those metals are driving their evolution. Scientists found that earthworms from contaminated sites can convert toxic metals into nontoxic materials, showing rapid adaptation to pollutants. In the case of arsenic, tolerant earthworms appear to produce a protein that binds the metal and keeps it inert.[14]

Scientists collected *Lumbricus rubellus* from lead and zinc mine sites in Wales and compared their tolerance to metal contamination with that of a population from uncontaminated soil. Not only did the worms from the mine sites accumulate more metal in their tissues, but they were able to convert a much higher proportion of the lead into a nontoxic insoluble material.[15] The genetic analysis revealed that the worms from the lead mine sites and the worms from the zinc mine sites came from two distinct genetic lineages, which means that the worms are rapidly adapting to all sorts of toxic materials. But is that good or bad?

Earthworms play a key role in soil formation and nutrient cycling—core ecological functions that support food production, climate regulation, flood mitigation, and water purification. Earthworms that can tolerate and neutralize heavy metals help sustain these functions and enhance soil remediation in contaminated sites. However, rapid adaptation to heavy metals can lead to disproportionate consequences for ecological community composition and biodiversity. Bioaccumulation of metals in earthworms poses risks up the food chain for ecosystems and human health.

The Heat-Tolerant Ant

City acorn ants (*Temnothorax curvispinosus*) can tolerate roughly 2°C more heat than their rural counterparts; they are also less tolerant to cold temperatures.[16] Using a common garden experiment, in which ants from different environments were reared under the same laboratory conditions, scientists have shown that evolutionary divergence (rather than plasticity) is responsible for the higher heat tolerance in urban acorn ants. Using ant colonies collected in field sites and controlling for environmental differences across populations, researchers can isolate the genetic component of the phenotypic divergence measured in the field. In an initial study, they found that urban acorn ant populations exhibit increased heat tolerance compared with rural populations. But only in a successive study were they able to disentangle whether the ant's tolerance was a plastic or a genetic response. A plastic response of the ant to the urban heat island would imply that the phenotypic differences observed in urban-versus-rural temperatures would disappear in offspring in the second generation. But that was not the case.

Cities' warmer temperatures are also responsible for earlier reproduction of urban acorn ants compared to nonurban ones.[17] While scientists have documented the effects of increasing temperature associated with climate change on the phenology of many plants and animals, only a few studies have focused on the implications of the urban heat islands for the timing of life cycle events directly related to fitness traits. Scientists were curious about the shift in timing of reproduction traits in urban acorn ant colonies as a result of urban warming,[18] and they found that urban ants reproduce approximately a month earlier than rural ants. This sounds like a minor detail in the timing of wildlife life cycles, except that this mismatch could limit the probability of cross-environment mating, leading to reproductive isolation between urban and rural populations.

The Metropolitan Rodent

The white-footed mouse (*Peromyscus leucopus*), a common resident of New York City's forest fragments, shows signs of directional selection in urban ecosystems.[19] Habitat fragmentation associated with the city's development over more than three centuries has led to geographic isolation and eventually the divergence of the mouse species into different evolutionary paths. Jason Munshi-South, then associate professor at Fordham University in New York, and his team examined 154,770 single nucleotide polymorphisms (SNPs) from 48 mice in urban and rural areas to identify urban adaptation. SNPs, genetic markers indicating variation at a single DNA position, enable the detection of population genetic variations. They discovered distinct selection signatures in urban populations, indicating adaptation to local selective pressures. The white-footed mouse exhibited positive selection in genes related to immune, toxicological, or metabolic functions, likely due to exposure to pollutants, physical barriers, and new foods. Each urban park within New York City hosted a unique mouse population, suggesting divergence due to the isolation of parks within urban environments.

Munshi-South and his colleagues also took tissue samples of the tail by trapping mice in New York City's remnant natural habitats. By sequencing the genome, they expect to unveil a suite of trait changes in city mice, such as a shift in the length of the mice's tailbones and limbs to match their styles of urban locomotion and a certain change in gut bacteria due to the new urban diet.

Munshi-South's lab team is also studying the brown rat (*Rattus norvegicus*), another New York City inhabitant. Brown rats show both adaptive resistance to the rodenticide commonly used in urban habitats and significant genetic differentiation within the urban environment, where high-traffic roadways and other artificial barriers limit gene flow.[20] By setting traps in New York subway platforms and storm drains, Munshi-South and his colleagues were able to determine that genes controlling the New York rat's olfactory sensors have evolved to help the rats navigate the city's new olfactory landscape.

The evolution of both mice and rats in the urban environment has important implications for human health. The white-footed mouse is a critical host for black-legged ticks, which carry and spread the bacterium that causes Lyme disease. Brown rats carry a number of pathogens and diseases, posing a significant risk for public health, especially among low-income urban residents living in poor housing conditions and neighborhoods with old or inadequate sanitary infrastructures.

The Fearless Bird

Songbirds in urban environments exhibit bolder behavior and lower levels of stress hormones.[21] Multiple stressors in cities challenge birds to develop coping mechanisms for living in this novel environment. Cities are bright and noisy, day and night, which disrupts the circadian rhythms of many animals. The multiple stressors are changing birds' endocrine mechanisms, selecting for endocrine traits that affect their ability to cope.[22] Stress hormones mediate the expression of behavioral traits and are associated with animal personalities.[23] Scientists have measured concentrations of glucocorticoid hormones in animal cells to establish whether there is a clear urban signal. The dark-eyed junco (a songbird), following its colonization in San Diego, California, was compared to the same species from a nearby wildland population in the species' breeding range; the scientists looked at levels of corticosterone in response to handling and flight-initiation distances (the buffer space a bird keeps between itself and an approaching threat) and established that urban dark-eyed juncos have lower levels of the hormone compared to their rural counterparts.

Corticosterone is a steroid hormone that plays an important role in the regulation of energy, immune reactions, and stress responses in many species. While elevated levels of corticosterone are often associated with adaptive short-term behavioral and physiological survival responses to external stress, chronically lower levels suggest adaptation to urban stress and are generally associated with bolder or more assertive behavioral phenotypes. However, the extent to which endocrine signals reveal a general urban pattern among birds and other organisms remains unclear.

The Shrinking Fish

River networks shape cities across the world and are, in turn, shaped by them. River valleys have historically been a major reason why cities emerge in specific locations. Rivers provided early urban settlements with a stable supply of water for irrigation and food production. Humans have modified rivers near cities for millennia, building large-scale levees, canals, and dikes to protect themselves against floods. Today's river networks embedded in the urban fabric are much different from the original geomorphological structures, and river networks do not provide the same function. Yet, streams and rivers still provide essential services to humans, including drinking water, irrigation, and hydropower.

Urban streams share characteristics that are distinct from their nonurban counterparts. Scientists refer to the physical and ecological responses caused by urbanization in these streams as the *urban stream syndrome*.[24] This syndrome includes significantly increased stream flows during rainfall because the land surrounding the urban stream doesn't absorb the water (as the land around a nonurban stream would). Rather, the water quickly runs off the land into the stream, producing short, intense flow pulses followed by a rapid decline in discharge—a hydrologic pattern known as a *flashy* hydrograph. The urban stream syndrome also includes elevated concentrations of nutrients and contaminants, altered channel morphology and stability, and reduced biodiversity.

While we know a great deal about how aquatic communities in urban streams respond to urbanization (e.g., their richness, abundance, and diversity), we know much less about how aquatic species respond evolutionarily to the environmental changes caused by urbanization. Only recently have evolutionary biologists started to study divergence in aquatic organisms in urban watersheds and coastal waters. The adaptation of the Atlantic killifish is one of the best-documented examples of evolutionary change in response to high levels of industrial pollutants, including dioxins, heavy metals, hydrocarbons, and other chemicals released into urban estuaries since the 1950s and 1960s. By sequencing the complete genomes of 384 Atlantic killifish from polluted and nonpolluted sites in Massachusetts, scientists were able to determine how the fish evolved tolerance to high levels of these toxins through changes in the aryl hydrocarbon receptor signaling pathway.[25]

Several scholars have documented the decline in fish size in response to the construction of dams and habitat changes.[26] They also show how natural selection in the body size of the Pacific salmon (*Oncorhynchus* spp.) affects salmon population dynamics, community interactions, and ecosystem processes (fluxes of salmon biomass).[27] Pacific salmon play an important role in the food web because they are a food source for many birds and mammals. Pacific salmon are also key to the cycling of nutrients by transporting energy and nutrients from the ocean into freshwater ecosystems.

New Species on the Block

> "Where do species come from?"
> "Where do they go when they go extinct?"

These are questions that Seattle first graders asked me on a field trip, during which I introduced them to urban ecology. Their curiosity about the fundamental processes of evolution was far beyond my expectations. They were not simply curious about why and how cities affect biodiversity and the implications of biodiversity loss for ecosystems and human well-being. They were also thinking about what makes the planet biodiverse, what processes generate new species, and what we can do to help that happen.

The first graders' question, "Where do species come from?" sounded very familiar. I told them that almost two centuries ago, an English naturalist, Charles Darwin, asked the same question while exploring the Galápagos Islands in the South Pacific. By closely observing birds and other organisms on different islands, he came up with an answer that changed the way we see the natural world today. He discovered that new species occur when a species splits into two or when a population diverges from its ancestor. Darwin was observing speciation, and his significant insight was that species evolved by natural selection, acting on variation among individuals.

If species are not created by magic but through evolution, and if cities drive species evolution, how do cities cause speciation? Divergent natural selection, in many taxa, can lead to speciation. Yet, species divergence does not necessarily cause speciation.

Among the few observed examples of urban-driven speciation is the London Underground mosquito. In the nineteenth century, a railway tunnel was sealed off, trapping underground the common house mosquito (*Culex pipiens*), which adapted to the environment and became genetically distinct from other mosquitoes living aboveground. The distinct underground *Culex pipiens* population was first observed during World War II, when German bombing transformed the underground tunnels into bomb shelters for more than 180,000 people. But this mosquito species is not limited to one city and one underground system. Since then, it has been reported in other urban underground infrastructure, such as water systems, sewers, and both the Tokyo and New York subway systems.

The evolutionary processes that give rise to speciation—including differential adaptation driven by divergent selection, divergence in mating traits due to sexual selection, genetic drift, and mutation—have all been described in relation to anthropogenic change and urbanization.[28] Yet we do not know how speciation is favored or hindered by a city's intense and multifarious selection pressures. What we know about the possible factors contributing to speciation, however, prompts a fundamental question about how urbanization affects the evolution of life on Earth. Do cities lead to speciation? And can the distinctive properties of urban landscapes unveil important insights about species extinction and speciation?

Through urbanization, humans mediate the interactions and feedbacks between evolution and ecology in subtle ways, such as by introducing changes in habitat structure and processes, altering landscape heterogeneity and connectivity, modifying biotic interactions, and introducing novel disturbance. By increasing social interactions, urbanization amplifies and accelerates environmental change. These diverse and co-occurring urban pressures on species' traits can increase both the total strength of selection on a given trait and the number of traits under selection, potentially leading to speciation.[29] Although there is good evidence that urbanization drives divergent selection, it remains unclear if urbanization also initiates speciation. This depends on whether divergence leads to reproductive isolation and if speciation is determined by genetic mechanisms.[30]

The way cities create distinct habitats above- and belowground is quite remarkable. The vast urban infrastructure creates distinctive habitat discontinuities that impose strong barriers to gene flow. Aboveground, tall buildings, highways, and distinct parks can separate populations within a species that typically exchange genetic material. Below the streets, the subways, railroads, tunnels, water pipes, sewers, drains, and power lines connect people and provide them access to places and resources while also separating what was previously connected in the ecological systems, both limiting and creating new pathways of dispersal. These underground structures are vital to the functioning of the city and to mitigating the impact of people on the preurban native ecosystem. The subway tunnel and storm drain, by breaking the continuity of nature or establishing new pathways, can dramatically alter species interactions, impede or promote gene flow, and potentially drive speciation.

The extent to which urban-driven population divergence leads to speciation may depend on the type of ecological shift in urban environments. Speciation can theoretically be promoted by divergence from extreme ecological shifts in one dimension or highly dimensional ecological shifts. However, it remains uncertain under which scenario speciation is completed, resulting in reproductive isolation.[31] Both the stronger selection hypothesis and the multifarious selection hypothesis could explain why cities might facilitate speciation.

Navigating Urban Evolution

Studying evolution in urban environments presents unique challenges and opportunities. In the first global study of its kind, our interdisciplinary team of evolutionary biologists and urban ecologists conducted a meta-analysis of over 1,600 documented phenotypic changes across diverse plant and animal species, ecosystems, and regions. We found that rates of phenotypic change—some confirmed to be evolutionary—were higher in urban ecosystems than in nonurban environments, including both natural habitats and other human-modified (nonurban) systems. Some of these changes reflect evolutionary shifts, though the study primarily targeted phenotypic traits, with confirmed genetic change included to a limited extent.[32]

A central question emerging from this work is whether cities leave a detectable and distinct evolutionary signature—one that can be distinguished from natural and other anthropogenic drivers—in the genomes of urban-dwelling species. The complexity of urban ecosystems, with their diverse, overlapping selection pressures and rapidly changing environments, requires new theoretical frameworks, methodological approaches, and cross-disciplinary collaboration. One of the key challenges is disentangling adaptive evolutionary change from phenotypic plasticity in traits that respond to urban pressures. Clarifying whether these changes reflect evolutionary adaptation or phenotypic plasticity is essential to understanding how urbanization shapes biodiversity.

How do we know that organisms are adapting to the urban environment? Urban-driven phenotypic change does not necessarily imply evolutionary change; instead, it can be plastic, meaning that the organism's phenotype is flexible and changes in response to different environments. Evolutionary change may also be nonadaptive, meaning that it is not driven by natural selection but rather the result of neutral processes such as mutation, genetic drift, and gene flow. To answer this question, we need evidence of both the trait's genetic heritability—the amount of phenotypic variation attributable to individual genetic differences—and their higher fitness in urban environments. Many experimental approaches (e.g., common garden experiments) are available and effective to test for local adaptation signals in species traits, yet they are logistically complex and not applicable to some species. Recent advances in genome sequencing and genotyping have made molecular approaches relatively more accessible, providing opportunities for combining approaches from population genomics with common garden experiments.

Urbanization affects natural selection by influencing fitness: an individual's contribution to future generations. Individuals with certain traits have higher survival and reproductive success rates than others and are therefore more likely to pass on these traits. Despite evidence of divergent phenotypic evolution in a wide diversity of traits (including life history, morphology, physiology, and behavior) between urban and nonurban environments, relatively few studies have linked genetic mechanisms with phenotypic evolution.

Perhaps the best-known case of human-induced evolution is that of the peppered moth (*Biston betularia*) in the 1800s. Industrial air pollution in Great Britain was shown to cause an increase in the frequency of the darker-colored moth.[33] A more recent example is San Diego populations of the dark-eyed junco, *Junco hyemalis*, which have less white in their tail feathers—a sexually selected trait—than juncos from mixed-coniferous temperate forests, likely due to urban selective pressures.[34] However, recent findings reveal inconsistent patterns across urban populations, including differences in tail length, wing length, and bill shape between cities, highlighting the complexity of these adaptations, which may be influenced by evolutionary histories, gene flow, environmental conditions, and selective pressures within and outside urban areas.[35] Among other examples of species for which there is strong evidence of genetic adaptation are the water flea (*Daphnia magna*), which has adapted to urban heat,[36] and the Atlantic killifish (*Fundulus heteroclitus*), which has adapted to high concentrations of polychlorinated biphenyls (PCBs) in four urban estuaries.[37]

Organisms are adapting to fragmentation (e.g., holy hawksbeard),[38] urban heat islands (e.g., acorn ants[39] and water fleas[40]), and chemical pollution (e.g., tomcod[41] and killifish,[42] both showing resistance to PCBs). Other studies of genetically based divergence in urban versus nonurban environments, including great tits,[43] white-footed mouse,[44] and crested anole lizards,[45] suggest evolution by natural selection although the evidence varies in strength across taxa, traits, and study designs.

Observations of trait changes in urban environments provide important clues about the mechanisms linking urbanization to evolutionary change. We are learning a great deal about how cities alter selective pressures and generate new ones, causing evolutionary shifts that alter ecological interactions.[46] Yet, we do not know the relative importance of different selective pressures, how these mechanisms interact and produce the observed genotypic and phenotypic shifts, and what traits are most likely to evolve. Most studies are explicitly designed to examine the effects of individual selective pressures rather than the potential interactions of multiple pressures and the extent to which their co-occurrence in urban environments might reinforce or constrain each other, leading to more stable or more unstable ecological conditions.

Because cities share many environmental features, they can be thought of as replicated experiments that provide the opportunity to explore and understand general patterns of evolutionary response in urban environments. However, cities vary in their morphologies, human population densities, socioeconomic structures, built infrastructures, and biophysical characteristics, which may lead to highly variable levels of disturbance and selective pressures.[47] Such environmental variation among cities may produce distinct evolutionary signatures. Only by accounting for such variability can we predict how organisms respond to urbanization and identify traits that enable them to thrive in different urban environments.

Detecting genetic signatures of adaptive and nonadaptive evolution across multiple cities is only a first step toward understanding how cities are shaping Earth's evolution by causing systemic changes in functional traits that regulate the productivity and stability of urban ecosystems. Intraspecific trait variation—variation in organisms' morphological, physiological, life-history, or behavioral characteristics stemming from genetic variability and phenotypic plasticity—has significant implications for ecological functions such as nutrient cycling and primary productivity. Integrating knowledge from all biological levels and time scales of evolution enables us to ask questions that we could not have imagined before and apply eco-evolutionary concepts to solve an array of socially relevant biological challenges, advancing our ability to predict potential urban eco-evolutionary outcomes under multiple urbanization scenarios.

Despite these complexities, urban environments offer unprecedented opportunities to observe evolution in action. Interdisciplinary studies combining evolutionary biology, ecology, urban planning, and the social sciences are essential for understanding the full scope of urban eco-evolutionary dynamics. Meta-analyses of phenotypic and genetic changes across species and regions, together with experimental and longitudinal studies, provide valuable data, revealing patterns and guiding future research.

The City as Evolutionary Lab

Urban environments have become the new laboratories for evolutionary biologists, with young scientists at the forefront of pioneering research. This new frontier presents opportunities and challenges as the assumptions of natural settings often do not match the reality of urban environments. In this section, we meet four young biologists who shared with me what sparked their scientific curiosity for studying cities, along with the unexpected challenges and surprises they've encountered in their research. Young scientists perceive the potential for their research to inform urban planning as a tremendous added value to their study of evolution in the city.

One Urban Garden at a Time

When Carly Ziter, now assistant professor at Concordia University in Montreal, Canada, joined Monica Turner's Landscape Ecology Lab at the

University of Wisconsin–Madison in 2014 as a PhD candidate, she chose an ambitious task: to investigate how landscape heterogeneity might affect the sustainability of ecosystem services across scales, from microhabitats to patches to landscapes to regions. But her exploration in this new landscape took a sudden turn (she calls it a "happy accident"). Exploring potential study sites in Madison, Ziter could not avoid encountering the city; what she called "nature" was now an urban landscape. Ziter saw an opportunity. "Urban ecosystems provide a new window into ecology," she says. Not finding a site that met her expectations presented a unique opportunity to ask her ecological questions in a new ecological system—one that challenges scholars trained in ecology. Many assumptions about representative sampling, study boundaries, and units of analysis—let alone collecting data—do not hold, and the literature offers limited guidance on what to expect.

Perhaps one of the most unexpected aspects of on-the-ground fieldwork in the city, compared with the countryside, is the presence of people. In studying nature in cities, the scientist must account for people; indeed, human behavior becomes a central element of the study. "While you could have predicted that," says Ziter, "what is a pleasant surprise is the direct experience of curious people that your presence in the field generates." People ask her a variety questions, and she says that they seem "genuinely curious" about the samples she collects and why she is collecting them—what she is measuring, what instruments she is using, and how those instruments work. They ask her what she's studying, why it's important, and what we can learn from the study. Ziter says, "People in the city are both the subject and the observer of [my] study." She was surprised by how excited city residents are about the science of the organisms that inhabit their own neighborhood and backyard.

One distinctive feature of urban landscapes is the great diversity of the patches that make up the urban mosaic.[48] Urban nature is not limited to larger parks; it is made up of fragments. It is in the myriad of backyards, rain gardens, bioswales, green roofs, street trees, and nature strips between the sidewalk and the street, as well as urban ponds, wetlands, and small creeks and rivers. All that we call green and blue infrastructure adds up to shape the ecosystem functions that support us and other species. What ecologists call environmental heterogeneity is amplified by the diversity of agents—households, businesses, and local governments—that produce the bits and pieces that make the urban landscape.[49] And these actions—such as developing a residential or commercial parcel, or landscaping and choosing what to plant in a backyard—are influenced by people's preferences and controlled through governmental regulations. When working in an urban landscape, "you realize that each small fragment of urban nature counts," says Ziter. This underscores the crucial role of each patch, garden, and backyard in supporting urban biodiversity and the power of urban agents in maintaining ecosystem resilience.

Life in the Urban Pond

Kristien Brans, now assistant professor at Vrije Universiteit Brussel in Belgium, has pioneered work on *Daphnia* in urban ponds, showing how urbanization drives evolutionary changes in life history, thermal tolerance, and physiological traits, all of which play important roles in ecological dynamics. When pursuing her PhD, she joined the Lab of Aquatic Ecology, Evolution, and Conservation at KU Leuven, directed by Luc De Meester, and she set out to test the evolutionary consequences of urbanization on *Daphnia* species through common garden experiments. Common garden experiments, Brans explains, "are experiments in which individuals of genetically distinct strains are reared under identical environmental conditions from an early life stage to disentangle the effects of genetic and environmental variation on the phenotype." This method, combined with advances in sequencing technology, provided a powerful approach to study local adaptation.

Brans's work shows evolutionary divergence of *Daphnia*'s life history, thermal tolerance, and physiological traits in urban populations. These traits play important roles in ecological dynamics and in eco-evolutionary feedbacks. For example, *Daphnia*'s grazing efficiency, which controls the formation of algae blooms in lakes and ponds, is correlated with the organism's body size. Genetic differentiation for functional traits in urban *Daphnia magna* may substantially influence ecological dynamics in city ponds and potentially modify grazing-associated ecosystem functions through eco-evolutionary feedback.[50]

One of her notable experiments involved exposing *Daphnia* populations to a simulated urban temperature gradient. They found a decrease in body size for *Daphnia* in ponds near built-up areas due to higher urban temperatures. However, in ponds with an abundance of large-bodied *Daphnia*, the average body size increased with urbanization, highlighting the complex impacts of urban environments.[51] Brans also discovered that urban *Daphnia* live a faster pace of life compared to their rural counterparts. Urban genotypes showed a faster lifestyle with reduced energy reserves and higher oxidative damage, and urban populations had a higher energy budget than rural ones, suggesting more efficient defense mechanisms and a reduced metabolic rate.[52]

Brans says that when the first data started to come in, it opened Pandora's box. "A lot is going on. . . . We also know that . . . phenotypic plasticity [is] involved."

Brans points out that conducting and replicating common garden experiments pose significant practical challenges, and such approaches may be possible only for certain species. Fieldwork in the city presents some interesting challenges and is an eye-opening experience for students trained in an evolutionary biology lab. Collecting thousands of samples across multiple sites and multiple cities in overpopulated regions, such as Flanders, Belgium, is time-consuming. To sample three urban ponds in three different cities daily, Brans's team had to carefully plan their schedule and take into account traffic jams. The field team would leave the lab at five o'clock in the morning and manage to return by seven o'clock at night, when the lab team would start to process the samples.

Brans says that some surprising findings and curious observations can only be experienced when working in a city. *Daphnia* live in small ponds and can even be found in fountains as small as two square meters. "Gene flow is human mediated. People are transferring the *Daphnia* population" to new environments based on how the ponds are managed, explains Brans. "They go through filters anywhere," and because workers often use the same device to drain multiple fountains, immigrant *Daphnia* can easily establish in a newly cleaned fountain. "And most striking," says Brans, is that in cities, "everything is changing very fast. An entire population [of *Daphnia*] can be wiped out due to the construction of an apartment building."

She emphasizes the potential to change how cities govern natural ecosystems through their findings, translating them into conservation planning. With advancing genomic methods and richer datasets, researchers have a unique opportunity to address critical questions in evolutionary biology, making significant strides in understanding urban evolutionary ecology.

Rats Take on the City

Jason Munshi-South, now professor in the Department of Biodiversity, Earth and Environmental Science at Drexel University in Philadelphia, did not expect to study evolution in New York City's parks. Trained as a tropical biologist, he started his career tracking elephants and parrots in the jungles of Africa and Borneo. But when he moved to New York City, he became fascinated by how small native mammals adapt to the fragments of green between skyscrapers and asphalt. It wasn't the rat that first fueled his curiosity, though; he first studied the white-footed mouse (*Peromyscus leucopus*), a native rodent. The white-footed mouse exhibited positive selection in genes related to immune, toxicological, or metabolic functions, likely due to exposure to pollutants, physical barriers, and new foods. Each urban park hosted a unique mouse population, suggesting divergence due to the isolation of parks within urban environments.[53]

This initial exploration of mice led to Munshi-South's genome study of rats. "We quickly realized that people were interested in rats. Urban dwellers deal with rats." Across the world, the rat population is increasing at a rapid pace in urban centers, and the recent explosion is made worse by climate change. Warmer temperatures mean rats breed faster. Brown rats (*Rattus norvegicus*) thrive in urban environments because they can rely on large amounts of human resources and waste.

In New York City, recent estimates suggest a population of two million rats lives underground, in subways and sewers, and aboveground in soft dirt. Rats are an increasing concern because they are major vectors and carriers of diseases. Mushi-South and his team set up a study to find out whether genetic adaptations play a role in the increase in rats. They collected samples and sequenced the genome of 29 brown rats from New York City and tested for genetic signatures of adaptation by comparing their genomes to the genomes of rats collected from Northeast China, the presumed ancestral range of the species. They found a series of candidate genes pointing to traits that have been subject to selection in city rats, including genes associated with diet, the nervous system, metabolism, locomotion, and behavior.[54]

Munshi-South explains that when he began studying the brown rat, he did not know the city very well, and available data were scarce. "I started driving, looking around, trapping rats for our study . . . [and] talking to people." He says he learned a lot from people he met while doing his fieldwork. "I learned that each part of the city, each park, has its own history. The cultural roots of the people vary." He heard stories from people, and through these stories, he "became aware that there is an inextricable relationship between culture, socioeconomics, and genomics." In addition, he says, "There is a lot of curiosity. People have a lot of questions. They want to know what function rats play in the urban ecosystem. And their questions prompt potentially interesting scientific inquiries." One question he heard is: "What will happen to the city as an ecosystem if we remove rats?"

Munshi-South also encountered new challenges in the field: "One [factor] is safety. The presence of people means also dealing with crime, drugs, [and] women and minority harassment. There are [also] issues of access, and variability of access, across different parts of the city. And then there is the bureaucracy that controls public land. Obtaining permissions [to conduct research] is not only frustrating, it also varies with neighborhoods according to different socioeconomic factors." And the urban landscape includes unique, three-dimensional space, along with sewers, tunnels, and subways. He explains: "When designing a sampling strategy, one should account for the many layers of the city, both in space and time."

Fearless Coyotes and Racoons

Chris Schell, assistant professor at the University of California, Berkeley, began his academic journey in psychology at Columbia University and rapidly discovered his passion for the evolution of animal behavior. He completed his PhD in evolutionary biology at the University of Chicago, where he explored the intersections of society, ecology, and evolution to understand how animals rapidly adapt to urban environments.

Schell says, "I became intrigued by urban carnivores: how society's perceptions and myths of these species had tangible consequences for their survival in cities. They were also extremely polarizing—some people loved them, while others were less enthused. My interests soon expanded to include how our perceptions and attitudes toward wild species shape our actions: how we manage the habitats animals reside in, policies that directly manage their populations, and what those actions mean for an animal's ability to adapt to urbanization. The mechanisms of transgenerational plasticity—through which parents are able to transmit their environmental experiences to their offspring—then became central to understanding how wildlife can quickly adapt to urban spaces."

In his work with coyote parents and offspring, Schell saw how central human–animal interactions were to his methods and questions. "Rather than making traditional remote observations of individuals in the field, we wanted to add a new wrinkle: what if we provided individuals with a food reward but simulated a 'threat' by remaining stationary in front of the animals? If we did this, we could assess how variation in boldness—a trait often implicated in urban wildlife—is directed toward the most influential organism in cities: people." Schell and his team noticed that "animal responses varied across individuals and with different people." Schell became interested in the cognitive processes that animals employ to assess risk and how their learning influences survival and/or reproductive success.[55]

In cities, many processes contribute to shaping phenotypic changes, leading to human–animal conflicts, which drive coevolution between humans and wildlife.[56] Studying animal cognition and human–animal interactions, Schell became curious about how human communities structure the urban environment—such as neighborhoods, buildings, green space, single-family and multifamily housing, private and commercial properties—and what strategies animals employ to navigate the urban environment. Schell wanted to know how animals (raccoons, specifically) learn to open boxes and trash cans—and what features of the urban environment enable their learning.

Not having a priori knowledge and expectations about the urban environment is unsettling for a young ecologist setting up an experiment in the field. "What surprises me the most," Schell says, "is how important and complex is the variable that we call *people* in a human-dominated landscape, a variable too often left out or inadequately represented in evolutionary studies. What drives the contrast among different landscapes cannot be fully understood by the change that the human footprint imposes on the landscape without understanding the socioecological dynamics that cause that change."

Schell continues: "Navigating the city to identify study sites, you can begin to observe differences among neighborhoods. It becomes apparent how the remarkable landscape differences that we see on the ground in land cover, land use, quality of the habitat, pollution, and access to green infrastructure have deep roots in social and economic inequalities, between industrial versus rich neighborhoods. You can observe the legacies that land use policies and regulations have left on the landscape. Redlining divided cities by race and income, relegating people of color to areas with limited access to resources. These differences can further explain the eco-evolutionary dynamics and feedbacks that we may detect within and across urban areas."[57]

As Schell reflects on the challenges ecologists face in studying the city, he says, "We are not trained to work in an urban environment, nor do we have connections to the fields that [possess] the different expertise that could inform a better understanding of the ecological and social interactions. Despite the progress that urban ecology has made towards interdisciplinarity, working across fields in a productive manner is still a huge challenge. We are still just learning how the fields intersect with each other."

A common thread among the young scientists I interviewed is their enthusiasm. Urban evolutionary ecology excites them because it emphasizes possibilities. This field turns the challenges of human–nature interactions into opportunities for environmental stewardship. Brans's eyes light up with excitement when she says, "We are just starting to uncover what drives urban evolution. What . . . we can find out can be applied in the real world." Young scientists see the emerging field of urban evolutionary ecology as both challenging and important.

Harnessing the City's Eco-Evolutionary Potential

It is well established that human activity causes evolutionary change. What makes the emergence and rapid development of cities across the globe a potential turning point in the evolution of Earth is the simultaneous and rapid change that urbanization imposes on multiple environmental and socioeconomic factors that drive evolution and their replication across the globe. By building cities, humans transform landscapes and microclimates, altering the connections between different habitats, changing the food web, influencing the composition of plant and animal populations, creating novel species interactions, and reshaping competition, predation, and symbiosis. These changes can lead to evolutionary shifts through their effects on natural selection, genetic drift, and gene flow.

In cities, we are unintentionally running experiments on how organisms respond to future environmental change. That is, today's cities and urban systems can be viewed as microcosms of the evolution of an urbanized Earth under different scenarios and thus offer an opportunity for designing global cross-city comparative studies to explore plausible eco-evolutionary futures of our planet. Understanding how urban environments alter evolution will help scientists develop models that can provide more accurate predictions, enabling more informed decisions to conserve evolutionary potential.

Over the past few decades, we have made remarkable progress in studying urban ecosystems and understanding how cities affect ecosystem processes and biodiversity.[58] However, we lack a clear understanding of the mechanisms that link urbanization to eco-evolutionary dynamics and the potential implications for urban resilience.[59]

Are there general traits likely to evolve in urban environments? How do development patterns create distinct landscape signatures and ecological changes that shape these traits? How do emerging urban phenotypes reshape the dynamics of populations, ecological communities, and ecosystems in ways that affect ecological and human well-being? Predicting evolution is one of the most fascinating and challenging problems in science. In complex ecological communities, biotic and abiotic factors can simultaneously influence evolution. A population can adapt to a new abiotic environment (e.g., a higher temperature), other species with which it interacts (e.g., a competitor), or a combination of abiotic and biotic factors.

Urban evolution has significant implications for urban planning and conservation. By recognizing cities as evolutionary environments, planners can design cities that support species and genetic diversity, thereby maintaining evolutionary potential. Strategies such as creating green corridors, preserving habitat fragments, and reducing pollution can enhance biodiversity and promote evolutionary resilience. However, the effectiveness of these interventions is tied to complex eco-evolutionary dynamics that are often disregarded in city design. By understanding and harnessing these dynamics, urban design and planning can better address the challenges of urbanization, helping cities contribute to maintaining global biodiversity. Integrating evolutionary biology with urban planning offers a powerful framework for creating sustainable, resilient urban ecosystems that benefit both humans and wildlife.

Cities, as dynamic and complex ecosystems, present unique opportunities to study evolution in action and apply these insights to urban planning and conservation. The future of urban evolutionary research lies in interdisciplinary collaboration, innovative methods, and a commitment to addressing complex challenges.

2 Rewiring Nature's Complexity

Throughout my childhood, the first cherry blossoms reliably marked the changing season; spring was announced with the emergence of butterflies, colorful blooms, and singing birds. Today, these rhythms have shifted. Since the 1950s, urban residents in the Northern Hemisphere experience spring and summer 1.6 and 2.5 days earlier per decade, respectively. And we are not the only ones noticing. This shift confuses bees with sudden flower abundance, disrupts plant pollination, and creates mismatches for birds like the American robin, affecting their food supply and offspring survival, and thereby impacting the entire ecosystem.

Urbanization and climate change create unique conditions that alter phenology—the seasonal timing of biological events like flowering, breeding, migration, and development. The anticipated timing of the warming season causes a cascade of eco-evolutionary feedbacks. What may seem a minor change in the timing of warming temperatures has the power to prompt evolutionary changes in the traits of many organisms. For instance, phenological mismatches between plants and pollinators may drive plants to evolve new colors, patterns, and nectar compositions to attract pollinators. The symbiotic relationship between plants and pollinators is just one of many biotic interactions affected by changes in seasonal timing. Temperature is the major abiotic factor that affects phenology. This subtle time change affects everything.

Cities reshape the evolutionary niches of organisms through various mechanisms: altering landscapes and microclimates, restructuring habitat connectivity, modifying food webs, changing species composition, and fostering novel species interactions that influence competition, predation, and symbiosis. Less recognized, however, are the subtle yet profound effects of urbanization. Artificial light disrupts nocturnal behaviors, altering feeding, mating, and migration patterns with cascading ecological impacts. Noise pollution interferes with animal communication, affecting mating calls and predator–prey relationships. Urban runoff introduces chemical pollutants into soil and water, accumulating in food chains and causing long-term health effects for wildlife and humans. Cities promote antibiotic resistance, as high concentrations of pharmaceuticals in wastewater drive the evolution of resistant bacterial strains. They also shape microbial evolution in ways that influence infectious disease dynamics—by concentrating hosts, altering species interactions, and creating conditions that can favor the emergence and transmission of pathogens. As hubs for invasive species, cities provide the disturbed habitats and abundant resources that enable these species to outcompete native species, disrupting local ecosystems and reducing biodiversity. Together, these pervasive factors fundamentally reshape ecological and evolutionary processes, revealing the intricate interplay between urbanization and natural systems.

Cities in Evolutionary Time

The eco-evolutionary changes we observe in cities today have precedents in earlier forms of human niche construction—but they are fundamentally different. Long before cities existed, humans were already reshaping ecosystems through socially transmitted practices, including the use of fire to clear land, intensive, species translocation, and selective cultivation.[1] By the Late Pleistocene, these activities had begun to restructure biodiversity, shift species distributions, and generate new ecological dynamics across entire landscapes.[2] Archaeological and paleoecological evidence shows that human niche construction transformed ecosystems at regional and continental scales thousands of years before industrialization. Sociocultural dynamics played a key role in shaping these long-term transformations, a point developed further in chapter 6.

Among the most consequential of these early transformations was the domestication of plants.[3] The agricultural revolution redirected the evolutionary trajectories of a large proportion of the planet's flora. Through sustained selection and ecological modification, humans altered plant morphology, reproductive cycles, and chemical compositions—engineering new traits to meet subsistence needs. These changes were far-reaching, but they were largely concentrated around a limited number of species and structured through relatively stable ecological relationships.

These early transformations unfolded over thousands of years through cumulative practices that modified ecosystems and species interactions across vast regions. Fire regimes altered successional dynamics and nutrient cycles.[4] Long-distance transport of animals and plants introduced new ecological assemblages. In some cases, sustained hunting pressure drove species to extinction or led to shifts in behavior and distribution that reverberated through trophic networks. Although these processes lacked the speed or infrastructure of modern development, they created enduring changes in the structure and function of ecosystems.[5] Many of these early transformations also produced lasting planetary-scale effects—from the extinction of megafauna and deforestation of entire regions to glacial retreat and oceanic changes associated with long-term greenhouse gas release.[6]

With the rise of permanent settlements, systematic environmental modifications intensified through expanded land conversion, irrigation, and species management. Early cities amplified these interventions, extending the spatial footprint and intensity of human niche construction, coordinating land use, water control, and species management at new scales. Still, it was the convergence of urban growth with industrial technologies that dramatically altered the pace, scale, and coupling of human and ecological processes. The integration of fossil energy, engineered infrastructure, and global trade and exchange reshaped not only how humans inhabited ecosystems but also how ecosystems evolved in response.[7]

Urbanization marks a turning point. It introduces a different set of drivers, acting across a broader range of taxa and ecological processes. Cities generate novel abiotic and biotic pressures: artificial light, chemical pollutants, habitat fragmentation, global species transport, and tightly coupled human-technological systems. These drivers operate with unprecedented speed, spatial reach, and connectivity. The changes they produce are qualitatively distinct. Through infrastructure, technology, social institutions, and cultural systems, cities establish new conditions for evolution—reshaping selection pressures across entire ecological communities.[8]

Situating urbanization within the longer arc of human–nature entanglement makes one thing clear: Cities are not separate from nature. They are embedded in deep evolutionary time. But they also mark a new phase—one in which human-built environments actively shape the evolutionary future of life on Earth. This shift reframes the central question. It is no longer whether humans affect evolution, but how cities—through their infrastructure, technologies, and coupled social-ecological systems—are rewriting the evolutionary play.

Cities as Evolutionary Agents

The satellite images of NASA's Visible Earth Night Light vividly capture the extraordinary human-driven transformation associated with urbanization, prompting us to consider cities' impact on Earth's evolution at a planetary scale. Humans in cities are not just altering local ecological conditions; they are fundamentally rewriting nature's rules. The challenge for urban ecologists and evolutionary biologists is to understand how cities shape eco-evolutionary dynamics.

Are cities changing Earth's evolution? Is there evidence of an urban signature of genetic change beyond the long-known natural and anthropogenic signals? How does urbanization drive evolutionary change? What species' traits are likely to evolve as a result of urbanization? What are the plausible consequences of these changes for ecosystems and human well-being? Can a new symbiotic relationship between humans and the planet—emerging through coevolution—change the course of each other's history?

Biology faces one of its greatest challenges since Darwin formulated the theory of evolution: the emergence of cities as novel agents of evolution. Humans are driving planetary change not only by altering the climate, disrupting biogeochemical cycles, and causing species extinctions—we are changing the genetic makeup of many organisms. While humans have been altering ecological processes for millennia, the convergent economic, social, and technological transitions associated with global urbanization have enabled a new capacity for niche construction—one in which organisms modify their environment and, in doing so, generate complex ecological and evolutionary feedbacks that shape the distribution and abundance of species.

By concentrating people, infrastructure, and technologies, cities fragment habitats, restructure ecological networks, and alter species interactions and abiotic conditions, giving rise to distinctive evolutionary niches.[9] The rapid evolutionary changes detected in individual populations within single cities may reflect processes operating at broader spatial and taxonomic scales. Although the evidence remains limited, synthetizing these observations and identifying the missing pieces of the puzzle are critical steps toward formulating hypotheses and designing experiments to test them.

The snapshots of real-time evolution described in the previous chapter prompt us to ask questions about the repeatability of evolution. Cities share many ecological features, such as elevated levels of impervious surfaces, increased temperatures, and high rates of species introductions, which make them real-time, large-scale experiments for determining whether urban-driven trait changes show convergence across species and/or parallel changes across populations of the same species. But cities also differ from each other in physical structure, human population density, social heterogeneity, and biophysical environments as a result of their histories, demographics, economic structures, and infrastructure. Thus, urban environments offer the chance to explore *which factors* drive parallel versus divergent evolutionary responses.

Comparable to archipelagos and oceanic islands, which, almost two centuries ago, provided a natural laboratory to understand biodiversity and evolution, cities provide a laboratory of contemporary evolutionary studies. Urban "archipelagos" prompt the same questions that led Darwin to formulate the theory of natural selection: What explains the uniquely high diversity of life and the variation among the islands in an archipelago? Darwin was witnessing evolution. His observations of how geological processes, geographic isolation, and colonization shape biodiversity laid the foundation of what ecologists Robert H. MacArthur and Eduard O. Wilson later synthesized into the theory of island biogeography.[10] Today, Darwin would probably ask similar questions: What explains the convergent patterns of diversity across urban environments that set cities apart from other habitats? And what explains the variations and inconsistencies among cities?

Urbanization is reshaping ecological and evolutionary processes, both locally and globally, with significant consequences for Earth's biodiversity. With global urbanization, humans have become powerful agents of natural selection, comparable to the geological processes that led to the geographic isolation and colonization of islands. Yet *how* these changes alter natural selection and affecting evolutionary mechanisms remains elusive. Since urban development creates a mosaic of natural fragments similar to habitat islands, the fundamental principles of island biogeography may apply to urban habitats. As with oceanic islands, the number of species in urban habitats reflects the balance between the rate at which new species colonize them and the rate at which populations of native species become extinct. Distant fragments of urban habitats will have lower immigration rates than those close to each other. Close urban islands will have high immigration rates and more species, and large islands will have more species than small ones due to their lower extinction rates.

However, urban islands differ from ocean islands. They emerge from complex interactions between humans and natural processes. Urban agglomerations create a myriad of landscape forms and compositions, and unlike geologic change, which is characterized by extremely slow processes punctuated by rapid events, urban change occurs at an exponential pace. In many ways, urbanization might be one of the fastest environmental changes, and therefore, it might swamp the adaptive responses of some species, replacing them with generalist, short-generation, or invasive species.

Shifting Species Coexistence

In nature, a multitude of interactions among species (competition, mutualism, predation, and symbiosis) support the diversity of life. These interactions form complex ecological networks that reflect the evolutionary history of their biotic components.[11] Mutualism between flowering plants and their pollinators is one of the most remarkable examples of coevolution. New evidence from DNA records suggests that bee diversification coincided with the largest flowering plant clade, the eudicots, during the Cretaceous.[12]

Predation, herbivory, and parasitism are examples of species coevolving through interactions that influence fitness. Predators and prey both evolve traits that enhance their ability to capture or evade capture. Plants have evolved chemical defenses against herbivores, and herbivores (such as butterflies) have evolved sophisticated chemical sensors that allow them to detect lethal compounds. Parasites tend not to kill their host since their survival depends on the host organism, yet parasites affect the host's fitness. As a result, host populations have been evolving resistance to parasites, which may, in turn, affect how hosts modify ecosystems.[13]

As species diversify—that is, as new species form and living species go extinct—interactions between species are added or removed, which alters community dynamics and ecosystem function.[14] The evolution of species changes ecological networks, which are also affected by evolving biotic interactions. Species do not live in isolation, and their interactions imply their reciprocal presence. As species disappear or as their traits evolve, this has consequences on other species and on their community. Coevolution in ecological networks helps us understand how urbanization reshapes species coexistence by rewiring network structure and composition, providing us with a key for predicting eco-evolutionary change in an urbanizing world.

Urbanization complicates ecological dynamics. Species vary greatly in their ability to adapt to urban environments based on dispersal ranges, generation times, genetic variation, and trophic positions; thus, urbanization does not affect all organisms equally. Different disturbances may have distinct effects for different organisms across space (from local environments to urbanizing regions) and time (including ecological and evolutionary time scales).[15]

The most apparent shifts caused by the urban rewiring of the natural world are in how birds, insects, amphibians, and other organisms move through urban environments and how—and how far—plants disperse their seeds. The built environment reshapes the dispersal pathway of animals and seeds by fragmenting the natural habitat, creating physical barriers, and, in some cases, even artificially connecting naturally isolated habitats. By altering dispersal, humans in cities change gene flow—the transfer of genes from one population to another. Gene flow can take place between two populations of the same species by transferring genetic variation from one population to another or between two different species. Gene flow increases genetic variation within populations and decreases it among populations.

Built environments typically act as a barrier that isolates different populations of the same species, which results in increased genetic drift within populations and reduced gene flow between populations.[16] However, human activities modify dispersal in complex ways, affecting gene flow, colonization, and community assembly. Humans reshape dispersal directly through the transport of organisms and indirectly by altering landscape structure, modifying dispersal vectors (both biotic vectors, such as organisms, and abiotic vectors, such as water and wind), and changing species dispersal traits.[17] These activities can hinder or facilitate movement among populations, leading to contrasting outcomes. Fragmentation may reduce genetic diversity within populations and increase differentiation between them. Yet human-built corridors may connect populations that were never previously connected, generating novel patterns of movement and gene flow.

What are less appreciated are the novel dispersal vectors and dispersal pathways that human activities create. Humans *intentionally* transport organisms when they use them in urban agriculture and landscaping, gardening, and ornamental planting. Humans in cities also *accidentally* act as a dispersal vector, for example, when seeds travel long distances on the wheels of cars or bikes. Bacteria, viruses, and parasites are unintentionally dispersed by human transportation systems. Urbanization increases intentional and accidental dispersal and alters species-specific dispersal likelihoods, distances, and destination.[18]

Biological dispersal—the movement of individuals from their natal sites to breeding sites and between breeding sites—depends on organisms' morphology, behavior, and physiology, as well as on biotic vectors (animals) and abiotic vectors (such as water and wind) that transport them. In cities, human-mediated dispersal creates unexpected patterns that are further shaped by the urban-altered landscape. The probability that individuals will disperse to a specific site and over a particular distance from their natal site (the dispersal kernel) is influenced by biotic and abiotic factors affecting departure from the natal location, of movement, and of settlement in the new location. Departure depends on factors such as local population density, resource competition, and habitat quality. Movement is influenced by landscape resistance. The propensity of organisms to settle in a particular site depends on environmental cues, such as population density or the availability of suitable mates at the new site.

Urbanization changes all these factors by rearranging the landscape structure and its biotic and abiotic features. Buildings, roads, and the extensive support infrastructure (both green and gray infrastructure) create new conditions that modulate and differentiate the propensity of different organisms to depart, the probability of their survival in the transfer, and the probability of successful establishment in the new location. Variable urban patterns and infrastructure across and within urban regions create a great variety of possible outcomes. Rapid evolutionary changes resulting from urban rewiring thus affect and are affected by dispersal in response to the changing environment.[19]

Complexity in Urban Ecosystems

A fundamental question in ecology is whether ecosystem complexity—defined by the diversity of species and the network of their interactions—is related to ecosystem stability, the capacity of the ecosystem to return to its functioning regime after disturbance. Ecosystems are structurally complex and exhibit complex behavior. An undisturbed tropical old-growth forest, with its multilayered canopy and diverse species interactions, can potentially recover from disturbances more effectively than a temperate second-growth forest subjected to clear-cut harvesting, due to the presence of multiple species that support diverse ecological functions. Urbanization is often held responsible for simplifying ecosystems by eliminating processes, landforms, species, and sources of ecological variation that form the basis for the eco-evolutionary processes that characterize healthy ecosystems. However, taking evolution into account, it is unclear how urbanization may alter ecosystem complexity and stability. The evolution of species' traits challenges the implicit assumption that the interactions among species remain static as adaptation can impact an ecosystem's response to perturbation.

Salmon and the Heatwave

Every year, salmon embark on a remarkable journey from the Pacific Ocean back to the streams and rivers where they were born, including those that run through urban areas like Seattle. This migration, critical for their life cycle, involves spawning in freshwater before the young salmon migrate to the ocean to grow and mature. In 2015, high temperatures devastated the sockeye salmon run in the Snake and Columbia rivers, which flow through the Pacific Northwest of the United States.[20] An estimated 250,000 sockeye died that year, long before reaching their spawning grounds. Similarly, daily summer temperatures in the Lower Green River, which flows through Seattle, typically reach 70–72°F and sometimes even exceed 74°F—a potential lethal temperature range for cold-water fish like salmon and trout. In

July 2015, temperatures exceeded this lethal threshold at almost every mainstem location sampled in the lower 45 miles of the Green-Duwamish River, the name given to the Green River as it flows through more urbanized stretches downstream. Scientists predict that by the 2080s, stream temperatures in the Puget Sound region will increase by 4.0–4.5°F due to rising air temperatures and decreased summer streamflow. Puget Sound rivers are expected to frequently exceed the thermal tolerances for cold-water fish species, significantly impacting adult salmon and trout.

This alarming trend adds to the multiple threats the Northwest salmon population has faced over the past century, including habitat destruction due to dam construction, overfishing, pollution from agricultural runoff, and urban development. The construction of dams has blocked migratory routes essential for spawning, while agricultural runoff has introduced harmful pollutants into their habitats. Rapid urban development has further encroached on critical spawning and rearing habitats, with significant implications for ecological stability.

The global warming of stream waters raises critical questions about how human alterations to river channels affect ecological resilience. Historical management practices, such as removing riparian shade and hydrological disconnection, make rivers vulnerable to warming. Simplified habitats lack functional redundancy—multiple species performing similar roles—essential for ecosystem resilience. Human actions, such as channel alterations and riparian land-use changes, reduce this redundancy, shifting communities toward more stress-tolerant species and increasing the vulnerability of degraded habitats to future temperature changes.

Urbanization has a profound impact on ecological complexity, as evidenced by the case of salmon populations in Washington State. The construction of dams has severely impeded salmon migration, spawning, and survival. The removal of some dams, such as the Elwha Dam and the Glines Canyon Dam in 2014, which were built without fish ladders, has demonstrated positive outcomes, enabling salmon to reoccupy their historical spawning grounds and boost their numbers. This underscores the significance of maintaining habitat connectivity for ecological resilience.[21]

Rapid evolution plays a critical role in the resilience of salmon populations to urban disturbances. As salmon face diverse environmental challenges, such as rising temperatures, habitat fragmentation, and pollution, their ability to adapt through evolutionary changes can significantly buffer the impacts of these disturbances. Adaptive traits, such as enhanced heat tolerance or modified migratory behaviors, can help salmon populations maintain stability in the face of urban-induced environmental changes.

However, the adaptive capacity of salmon is influenced by several factors, including genetic diversity and environmental pressures. Urbanization can lead to population bottlenecks and reduced genetic diversity, constraining salmon's ability to respond to future environmental changes. This loss of adaptive potential can significantly hinder the recovery of salmon populations even after the removal of dams or other barriers. Without sufficient genetic diversity, the range of adaptive responses available to a population is reduced, increasing its vulnerability to new stressors. Therefore, while rapid evolution can provide short-term resilience, it may also pose long-term challenges for the recovery and stability of salmon populations in urbanized ecosystems. Maintaining and enhancing genetic diversity within salmon populations is crucial for ensuring their long-term resilience and capacity to adapt to ongoing and future environmental changes.

Understanding how urban-driven evolution affects ecological complexity necessitates an expanded concept of resilience and stability. Stability means not only maintaining processes despite environmental fluctuations but also evolving to support new functions. Living systems require both robustness (resilience to environmental changes) and flexibility (the ability to adapt to new challenges). Maintaining essential traits and fostering the capacity to innovate in response to environmental shifts are vital for survival. While restoring rivers to enhance connectivity and habitat diversity is crucial for building resilience against various pressures, including climate change, maintaining ecological stability requires moving beyond a simple return to preurban conditions. It involves acknowledging evolutionary dynamics and aiming to maintain the adaptive potential of species to support their long-term survival and ecosystem health.

Opportunities for Innovation

Urban ecosystems, as complex systems coupling humans and nature, are characterized by their hybrid nature, which is both unpredictable and innovative, fostering coevolution and adaptation. Consider interspecific hybridization in genetics, in which the interaction of different genomes creates novel phenotypes. This concept can also apply to urban systems, which maintain existing functions while adapting to new circumstances.

In *Robustness and Evolvability in Living Systems*,[22] evolutionary biologist Andreas Wagner argues that robustness allows biological systems to maintain function across diverse genetic configurations, enabling evolutionary change to accumulate genetic differences while preserving core functions across a range of conditions. For Wagner, robustness does not arise simply from redundancy, but from the organization of biological networks in which multiple genetic or regulatory configurations can produce the same functional outcome. This property creates the conditions under which evolutionary novelty can emerge when environments change.

Hybridization provides a powerful context in which this latent potential can be expressed, as genomes that have evolved separately are brought together and interact in new ways.[23] Interactions among divergent regulatory and coding elements can generate phenotypic outcomes that are not predictable from either parental lineage alone, sometimes revealing cryptic variation and sometimes producing incompatibilities. Modeling studies and empirical syntheses suggest that hybrid populations may, under certain conditions, adapt rapidly to novel environments, with fitness advantages that can persist across generations and are not strictly determined by ploidy. At the same time, these same dynamics can threaten biodiversity when extensive hybridization leads to introgression or the erosion of isolated lineages.

In urban ecosystems, opportunities for resilience and innovation arise from the hybridization of historical and novel structures, retaining historical functions despite rapid environmental change. Robustness in this context refers to a system's ability to absorb disturbance by reorganizing interactions—shifting among alternative strategies, such as resource partitioning, altered competitive hierarchies, or coexistence with newly introduced species. Urban food webs often exhibit increased interaction diversity, which can enhance robustness, even when species richness remains unchanged. Crucially, the consequences of species loss depend on a species' position within the network rather than on richness alone. The removal of a top predator, such as wolves, can disproportionately destabilize urban-adjacent ecosystems by releasing prey populations from control, triggering cascading effects on vegetation structure and associated species.

Urbanization influences the balance between adaptive evolution and species sorting (matching species to suitable environments). It might reduce colonization rates, making it difficult for new species to settle in isolated urban areas. Without frequent colonizations, resident species may monopolize resources, hindering new species from establishing. For example, the gray squirrel's dominance in urban parks can prevent other species from thriving. Conversely, increased connectivity in urban areas can promote species sorting, allowing species to find optimal environments by shifting ranges. Birds like the peregrine falcon, which have adapted to nesting on skyscrapers, illustrate this adaptive range shift.

Evolutionary feedbacks vary with species interactions, influencing how evolution and ecology shape species diversity through competition, predation, or mutualism. Environmental changes can drive species adaptation, altering these interactions. However, evolution does not always enhance the stability and diversity of ecological networks. Phenological asynchrony, in which climate change decouples the timing of life events, can disrupt interactions between insect herbivores and host plants or flowering plants and pollinators. Earlier flowering driven by warming, without corresponding shifts in pollinator activity, creates phenological mismatches that undermine pollination and affect both plants and pollinators.

Insect herbivores have coevolved with plants for millions of years, but human activities over the past 12,000 years have significantly altered these relationships.[24] Climate change shifts the range and phenology of insect herbivores, affecting their food sources and evolutionary potential. For example, as temperatures rise, some butterflies emerge earlier, but their host plants may not be available, leading to population declines. Conversely, plants with high evolutionary potential might escape predation pressures, leading to population growth. In mutualistic relationships, both plants and pollinators can adapt their timing in response to climate change, but rapid adaptation without corresponding changes in mutualistic partners can reduce fitness and limit evolutionary rescue. This scenario is evident in the declining populations of bees, which struggle to adapt to rapid changes in flowering times.

The complexity of urban ecosystems, characterized by diverse species interactions, makes predicting the impact of urbanization on ecological network stability, biodiversity, and ecosystem function challenging. However, by examining specific examples and understanding the underlying principles, we can better appreciate the opportunities and challenges that urban environments present for innovation and adaptation.

Emergence of Hybrid Networks

As in biological systems, human communities depend on complex networks to distribute energy, materials, and information. Global urbanization reshapes both human and ecological communities by redefining network structures and interactions.[25] In cities, social, built, and ecological networks coexist, often operating according to different logics and on different spatial and temporal scales. Often these networks intertwine, creating hybrid networks that support hybrid functions. Furthermore, urban infrastructure drives species evolution and the emergence of new networks.

By rewiring connectivity among habitats, populations, and communities, urban infrastructure can shape species dispersal strategies, alter species composition, and restructure food webs, with consequences for ecosystem function and trait evolution. Hybrid networks such as green corridors or linear parks simultaneously support human and ecological functions. Green corridors in cities are designed to provide habitat for wildlife while simultaneously mitigating stormwater flows, reducing heat, cooling the atmosphere, and absorbing atmospheric pollutants. Green infrastructure supports human health and provides opportunities for recreation[26] and ecosystem stewardship.[27]

While hybrid networks can enhance connectivity and multifunctionality, their ecological consequences depend critically on how species interact within them. Species interactions shape population dynamics across spatial and temporal scales, yet they are often overlooked in assessments of urban ecological change. Studies of interaction networks reveal how urbanization reorganizes ecological relationships in ways that affect ecosystem stability and function. By studying interactions between plants and nesting birds across an urbanization gradient, Amanda Rodewald and Stanley Gehrt showed that as landscapes urbanize and human population density increases, bird–plant networks become more nested, less compartmentalized, and increasingly dominated by strong interactions among few species.[28]

These changes in network structure in urban environments reduce avian nest survival by simplifying nesting environments and increasing the search efficiency of predators. Interestingly, Rodewald and Gehrt also describe what they call the "predator's paradox": A key factor of success for the predator is no longer the number of predators per nesting birds (known to be a major factor in urban environments) but instead the altered trophic dynamics and food web interactions. Urbanization reduces top-down control by increasing bottom-up mechanisms through the greater availability of anthropogenic foods, which support high densities of generalist predators and shift foraging behavior toward reliance on anthropogenic resources. Thus, it is the reorganization of interaction networks that plays a central role in shaping ecological and evolutionary outcomes in cities.

Just as new synapses form in our brain, giving rise to complex neural networks to exchange information, in biological and social systems, novel connections are continually created, and those no longer in use are lost. In urban ecosystems, a novel form of cooperation may emerge between human and natural systems to support what I define as "hybrid" functions. For example, the creation and maintenance of multifunctional green space in urban ecosystems require the cooperation of multiple agents and must also synchronize with the natural processes and organisms that govern them.

Urbanization thus profoundly rewires connectivity among habitats, populations, and communities. These changes affect species with different dispersal strategies and can lead to the evolution of dispersal.[29] Through changes in species composition and altered food webs, rewired connectivity networks affect ecosystem function and feed back on trait evolution. Mapping eco-evolutionary feedbacks and their implications for our species on an urban planet will require rethinking both ecosystem science and evolutionary dynamics. Humans are moving ecological systems beyond the envelope of conditions that have existed throughout evolutionary history.[30] We are changing what G. Evelyn Hutchinson, the father of modern ecology, described as the ecological stage in which the evolutionary play unfolds.[31] By changing the ecological theater, cities are writing a new eco-evolutionary play.

The Urban Eco-Evolutionary Play

Humanity has become a dominant force shaping Earth's future. Over the last 300 years, human societies have reshaped Earth systems at a scale comparable to major geophysical forces.[32] Rapid growth in population, energy use, and material throughput has altered the movement of elements across the planet, intensifying carbon, nitrogen, and phosphorus fluxes with cascading consequences for climate regulation, ecosystems, and human well-being. Large-scale infrastructure has reconfigured hydrological connectivity, transforming river networks through dams, reservoirs, and diversions, while land conversion and biomass appropriation have redirected a substantial share of global primary production toward human use. These transformations accelerated with industrialization and intensified in the twentieth century as energy availability, technological capacity, and economic activity became tightly coupled.[33] The resulting shift in planetary-scale dynamics has set a new context in which ecological and evolutionary processes increasingly unfold within human-dominated systems. Global urbanization may represent a turning point in this trajectory by concentrating these processes in space and time.

The Earth's eco-evolutionary play is a coauthored show. Its actors cocreate the performance. Contrary to a theatrical production, there is no producer, director, or playwright. The multitude of living organisms—the actors—are in charge. They participate in creating the conditions for evolutionary change, and their traits are the object of evolution. The actors' performance is shaped by environmental constraints and opportunities, interactions among actors, and their capacity to adapt. Urbanization is transforming the ecological stage—eliminating and introducing new actors, as well as reshaping the adaptive landscape. This new eco-evolutionary play, in turn, transforms the theater by altering both the actors and the stage.

Evolutionary biologists Kevin Lala (previously Laland) and Lynn Chiu use Sewall Wright's influential visual heuristic—the adaptive landscape—to illustrate how evolution takes place if we view organisms as key actors in the evolutionary play. The adaptive landscape is a surface that shows, in multidimensional space, the relationship between the fitness of a population and one or several traits or genes. The mean fitness of a population (its reproductive success) evolves as allele frequencies change. Evolution by natural selection—the change in the genetic makeup of populations—can be described using the adaptive landscape as a computational process that solves an optimization problem through an iterative search algorithm, known as a hill-climbing: Populations "climb up" to the "adaptive peaks," the allele combinations with the highest fitness.

Niche construction theory proposes that organisms influence their own evolution and the evolution of other organisms, both through natural selection and as agents of the conditions that shape that selection. Constructed environments and their legacies drive evolutionary processes. Therefore, organisms are both the subjects and the drivers of evolution. By transforming the environment, organisms can determine the conditions for their reproductive success. Species and their environments are intertwined in a continual feedback loop, leading to ecological and evolutionary change. Neither the species nor the environment is static. These dynamic interactions occur not only over longer time scales; they may unfold over just a few generations. This is what makes eco-evolutionary feedback remarkable.

Humans as Planetary Engineers

All living organisms shape their surroundings to enhance survival, and humans are no exception. In his later years, Darwin closely studied the role of earthworms in soil development. By consuming plant matter and soil, earthworms enrich their casts with concentrated organic and mineral content, making them more nutrient-rich than the surrounding soil. Other species, such as beavers, modify entire landscapes by building dams that restructure hydrology, increase habitat complexity, and influence species diversity across riparian systems. Ecologists refer to such organisms as ecosystem engineers—organisms whose environmental modifications significantly affect the abundance, distribution, and interactions of other species, as well as ecosystem functions at local and broader scales.

So what makes humans different? What made humans a dominant force on Earth? According to Lala, what distinguishes humans is their capacity to acquire, share, and store information at unprecedented scales, enabling cultural knowledge to accumulate across generations.[34] In other words, it is our complex culture. It was Darwin who first pointed to the role of culture in evolution. (However, the Modern Synthesis—the early twentieth-century unified theory of evolution—ignored his insights on culture, focusing exclusively on his insights on genetic evolution and natural selection.) The emphasis that Darwin placed on the role of agency and culture is particularly significant in light of the increasing evidence about how culture drives human evolution. We have good evidence that social learning and culture are not exclusive to humans. What makes culture different in humans is the scale and method of transmission. Learning and cumulative culture, aided by sophisticated language in humans, allow for the rapid development and diffusion of adaptive innovations to face environmental challenges. And cultural evolution may occur faster than genetic evolution via natural selection. According to niche construction theory, gene–culture coevolution may drive genetic evolution in the same way that genetic evolution can bias cultural evolution.

Why did our species and not others evolve special abilities for cumulative cultural learning? And why did it emerge during the Pleistocene rather than in earlier stages of human evolution? Learning and cooperation are two key traits of humans that might explain the emergence of cumulative culture. Peter Richerson and Robert Boyd suggest that "human life history is adapted to exploit the adaptive advantages of culture."[35] Compared with other apes, humans reproduce on shorter timescales, while also occupying a much broader geographic range and maintaining extensive social networks. Together, these features increased the importance of socially transmitted knowledge for coping with environmental variability. During the Pleistocene, climate-driven changes amplified spatial heterogeneity across landscapes, creating conditions in which flexible learning and cumulative cultural practices became increasingly advantageous. These pressures likely favored the evolution of enhanced learning capacities and cognition, with downstream consequences for brain expansion.

One factor that may help explain the rapid pace of human evolution is the emergence of extensive cooperation, including forms of cooperative breeding in which individuals beyond parents contribute to raising offspring. Ethnographic studies of hunter-gatherer societies document high levels of nonkin cooperation within and across groups, creating social contexts in which knowledge and skills can circulate broadly. Analyses of residential patterns across multiple foraging societies reveal a distinctive large-scale social organization—characterized by flexible group membership and repeated interaction beyond immediate kin—that is not observed in other primates.[36] Such social structures expand opportunities for social learning and help explain how cumulative culture could emerge and persist. In contemporary societies, cities intensify these same dynamics by concentrating social interaction, making visible how human culture and cooperation have reshaped evolutionary processes and extended niche construction to the planetary scale.

The transition from hunter-gatherer society to farming 12,000 years ago and the emergence of the first urban settlement 5,000 years later may have been turning points driven by the evolution, in size and structure, of the social networks that supported human societies. This transition marks the emergence of the technosphere—a global system of human-made technologies and infrastructures—that has vastly amplified human environmental agency.

Urbanization represents a key transition of human niche construction. While cities have been around for millennia, it is only over the past century that the human species has become an urban species. And with urbanization, humans' extraordinary capacity to develop and use technology has greatly expanded the boundaries of their realized niche, turning our species into planetary engineers. Urbanization is both a result of the evolution of human niche construction and a driver of the economic, social, and technological transitions that have expanded and accelerated human capacity for niche construction, generating complex feedbacks in both ecological and evolutionary processes. Urbanization is a shift in the dynamics of social interactions among people and has the potential to change the interactions between people and other species.

Reimagining Urban Nature

Understanding the role of humans in the eco-evolutionary play requires reimagining our place in nature. The long-held narrative that humans can control nature has shaped human identity and imagination, extending even to how we envision alien worlds, where the same dynamics of dominance and control are often projected. Evidence points to a different story. Humans do not control nature, nor are they independent of its outcomes. Human agency is inseparable from ecological and evolutionary dynamics. Our capacity to transform the planet marks a novel phase in Earth's evolutionary history—one in which humans may influence their own fate, but not determine the fate of the planet.

One example of the interplay between natural and human action is the relationship between climate change and hydropower. The Mekong Basin is one of the richest areas of biodiversity in the world. In November 2012, Laos announced the construction of the Xayaburi Dam on the Mekong River, despite opposition from Cambodia and Vietnam, which were concerned about its impact on the livelihoods of millions of people downstream who depend on the river. The dam would generate 1,260 megawatts of energy, and 95% of the power would be exported to Thailand. Eleven dams in China and two in Laos have altered the Mekong River's geomorphology and hydrology and reshaped its ecological dynamic, redefining its relationship with the human communities that depend on it.

One of the longest droughts ever experienced by the Mekong region has resulted in the lowest water level in a hundred years; parts of the river are no longer a waterway, putting at risk one of the world's largest inland fisheries, which provides food security and livelihoods for 60 million people living downstream in Cambodia, Laos, Thailand, and Vietnam. The lower Mekong River basin supports the largest freshwater fishery in the world, with over 2 million tons of fish harvested annually. As the future of the 2,700-mile river that has nourished people for millennia, is increasingly uncertain, we ask whether nature will reverse the decision to build the Xayaburi Dam and many other dams.

Will the expected droughts and the ecological and social threats force the people in charge to go back to the drawing board? It might take a long time, but as the ecosystem loses resilience—its capacity to adapt to change—greater are its chances of reaching a tipping point. Together, hydropower, climate change, and habitat loss will inevitably change the future of the people who depend on the river. The changing ecosystems will shape the choices the community faces: to prevent, adapt, or respond to the consequences.

A positive feedback between human agency and ecological dynamics could prompt evolutionary changes and the emergence of innovations in human–nature cooperation. Social groups with advanced social systems may be subject to cultural group selection for their advanced capacity to adapt and for their nature-oriented social norms and institutions. As in the Pleistocene, when increasing climate variation might have favored the evolution of social learning through the evolution of toolmaking and the development of large brains, the recent acceleration of environmental changes may prompt a new level of cooperation between our technological species and the biosphere.

Studying how humans mediate eco-evolutionary dynamics through urbanization can contribute to a new synthesis in evolution and ecology. By integrating humans into the study of eco-evolutionary feedbacks, ecological scholars might be able to reconcile key concepts, including niche construction and community assembly, and redefine Hutchinson's "realized niche" in an urbanizing planet. Humans have transformed both their fundamental and realized niches through technology, controlling environmental conditions and interactions with other species. This transformation affects niche construction of other species by modifying habitats, predation, and resource variability. Urbanization marks a key transition in human niche construction. While cities have existed for millennia, only in the past century have humans become predominantly urban, expanding their realized niche and becoming planetary engineers.

Understanding evolution in the presence of humans reveals our capacity to choose bio-cooperation. Rethinking evolutionary processes in a human-dominated world also expands the notion of evolutionary causes to include phenotypic plasticity and social learning. The central intellectual challenge for both the human and natural sciences—and for those in society actively engaged in shaping Earth's future—is to move beyond simplified narratives of catastrophe versus optimism. Can we imagine a future where humans have coevolved with other organisms? Can new symbiotic relationships emerge in which human actions, in cooperation with other species, create conditions that sustain life? What would such a world look like? What kinds of cities will we build and inhabit? Drawing on empirical observations and emerging theory, I outline the attributes of what I call bio-cooperative cities—cities that embed a new cooperative relationship between humans and other organisms to sustain ecosystem function and human well-being over the long term.

3 Symbiosis in the City

Microbes are the smallest inhabitants of cities, yet they are essential to the biological and ecological processes that sustain urban systems. While known primarily as pathogens and potential threats to human health, microbial symbioses play a key role in maintaining major ecological functions that directly support humans and city life. Mycorrhizal fungi form symbiotic relationships with plant roots, enhancing water and nutrient uptake—functions essential to the success of urban green roofs and parks. Symbiotic nitrogen-fixing bacteria in the roots of leguminous plants similarly improve soil fertility in urban community gardens, reducing the need for chemical fertilizers. Since these symbioses are invisible to humans, we tend to underestimate their importance in maintaining ecological and human well-being.

Beyond microbes, symbiotic relationships in cities span a wide array of organisms and functions. Lichens—formed through partnerships between fungi and photosynthetic algae or cyanobacteria—colonize walls and sidewalks, tolerating pollution while stabilizing surfaces and contributing to nutrient cycles. Pollinators like bees and butterflies engage in mutualistic interactions with flowering plants in parks and gardens, enabling reproduction and food production. Birds and fruit-bearing trees exchange services through seed dispersal and nutrition. In engineered systems, such as green infrastructure or stormwater wetlands, symbioses between aquatic plants and microbial communities facilitate water purification. These diverse associations among organisms actively sustain and sometimes reconfigure ecological functions in the heart of the city.

Symbiosis refers to the close and sustained interaction between different species. These relationships span a spectrum—from mutualism, where both partners benefit, to commensalism, where one benefits without affecting the other, to parasitism, where one benefits at the expense of the other. Symbiotic relationships evolve, dissolve, and reconfigure over time, often in response to environmental change. Some symbioses are obligate, meaning the participating species cannot survive without each other; others are facultative, offering advantages without being essential. In all cases, symbiotic relationships shape how organisms function, adapt, and persist within their environments. In urban systems, new interactions may be formed or existing ones disrupted as species encounter novel partners, altered habitats, and unfamiliar stressors. Understanding the range and plasticity of symbiotic relationships is essential for recognizing how cities might host and generate new forms of interdependence, as well as harness them to support resilience and adaptability.

In my recent summer class on urban ecology and evolution, my students and I explored the potential of integrating knowledge about nature's symbiotic relationships into city planning and green and blue infrastructure. We examined how rapid urbanization could reconfigure symbiotic relationships that have evolved over millennia and the potential for new symbiotic relationships to emerge and enhance urban resilience. I emphasized the need to consider the intricate interplay of social, ecological, and technological facets of urban systems and their eco-evolutionary dynamics. Overlooking these factors can undermine the effectiveness of these interventions, leading to unforeseen consequences such as species evolving resistance, unintended evolutionary trade-offs destabilizing ecosystems, or disruptions in the species interactions that these solutions aim to preserve. Thus, a thorough understanding of eco-evolutionary dynamics is crucial to harnessing the full potential of symbiotic relationships in urban environments.

Symbiosis as a Source of Evolutionary Innovation

Symbiotic interactions are important drivers of biological diversity and evolutionary innovation, though their role as sources of evolutionary novelty has only relatively recently been recognized. In 1905, Russian botanist Konstantin Mereschkowski hypothesized that eukaryotic cells (cells with membrane-bound organelles, including a nucleus) evolved through the symbiosis of single-celled prokaryotes. However, it wasn't until 1967 that evolutionary biologist Lynn Margulis provided microbiological evidence, and in 1978, Robert Schwartz and Margaret Dayhoff established the first experimental evidence. Margulis's theory of endosymbiosis profoundly transformed our understanding of the evolution of eukaryotic cells (which include all plants, animals, and fungi) through partnerships involving various prokaryotic cells (including bacteria and archaea).

"Life did not take over the world by combat, but by networking," wrote evolutionary biologist Lynn Margulis, capturing the essential role of cooperation as an evolutionary force. Her studies on cell structure highlighted symbiosis as a major driver of evolutionary change.[1] From the emergence of eukaryotic cells—formed through the merging of separate prokaryotic organisms—to the rise of multicellularity and complex eusocial societies, cooperation has catalyzed major evolutionary transitions. While competition has long been a cornerstone of evolutionary theory, it alone cannot account for the complexity and innovation inherent in living systems.

Recent expansions of evolutionary theory now acknowledge cooperation not as an anomaly but as a fundamental driver of biological organization. Cooperative relationships—whether among genes, cells, individuals, or species—can lead to new forms of integration and emergent function. These transitions, such as the merging of once-independent organisms into new wholes, suggest that evolutionary change often depends as much on alliance as on competition. This broader view opens new ways to understand how mutualisms emerge, stabilize, and shape the trajectory of life, particularly in environments that demand flexibility, innovation, and shared resilience.

Facilitation and cooperation emerge as critical strategies, particularly under environmental stress, enabling the rise of new organizational hierarchies and levels of complexity. As ecosystems become more urbanized and urban environments intensify in complexity, imposing novel stressors, the capacity for symbiosis to drive evolutionary novelty may become increasingly important. Urban environments act as powerful arenas for testing the evolutionary resilience of cooperation. Whether through microbial communities adapting to chemical stress, pollinators modifying behavior in tandem with microbes, or new associations forming in abandoned lots, cities offer a landscape where cooperation must not only persist but can also evolve.

Symbiosis is an evolutionary strategy that enables species to adapt to diverse habitats. In a rapidly changing world, the survival of many organisms depends on the symbiotic relationships they can form and maintain. Symbiosis grants organisms novel functions and the ability to expand into new ecological niches. Emerging feedbacks between ecological and evolutionary processes play a fundamental role in the emergence and dynamics of species interactions (both competitive and symbiotic), driving biodiversity and maintaining ecosystem function. In diverse natural communities, species interact in complex ways that alter community composition and trait selection.

By grounding symbiotic innovation in evolutionary theory, we better understand the underlying mechanisms that allow symbioses to emerge, stabilize, and transform ecosystems—even in the most human-altered environments.

Symbiosis as an Urban Evolutionary Strategy

Cities concentrate diverse species in dense, rapidly changing environments. These conditions intensify competition and open new possibilities for cooperation. While urban evolution research has predominantly focused on competitive dynamics and abiotic stress, the evolution of symbiotic relationships—mutualism, commensalism, and parasitism—remains underexplored. Yet symbiosis offers powerful strategies for adaptation: buffering disturbance, accessing new niches, and stabilizing ecological functions under stress. In urban ecosystems, these partnerships act as engines of evolutionary innovation.

Recent scholarship has emphasized the importance of cooperative interactions in biological systems. This book builds on that foundation by framing symbiosis as an evolutionary strategy shaped by urbanization and rapid environmental change.

Human activities reshape habitats, filter species assemblages, and expose organisms to novel selective pressures that can disrupt existing symbioses or catalyze new ones. As species navigate fragmented habitats, altered resource flows, and unfamiliar communities, survival may depend on the capacity to form adaptive associations. Some symbioses erode as partners lose compatibility. Others emerge where overlapping ecological needs foster functional complementarity—on green roofs, in stormwater systems, or within remnant patches embedded in the built environment.

Urban heat islands shift the timing and composition of insect–plant mutualisms. Pollinators adjust foraging behavior and partner choice in response to fragmented floral resources. Mycorrhizal associations in compacted or nutrient-altered soils may change composition or function. Some species lose mutualists due to pollution or fragmentation; others gain novel ones, including domesticated or introduced species. As urban assemblages reorganize, they produce interaction networks rarely seen outside cities.

Built environments also generate conditions for novel symbioses. Green roofs, bioswales, and restored corridors function as dynamic habitats where plants, microbes, fungi, and invertebrates form multispecies networks. Stormwater systems support aquatic plant–microbial partnerships that cycle nutrients and filter pollutants. Wastewater treatment depends on microbial consortia performing key ecological processes. These infrastructures both sustain existing relationships and foster new ones shaped by selective pressures favoring interdependence.

Some symbioses evolve in response to persistent urban selection. Mutualistic seed dispersers shift behavior in response to artificial food sources. Microbial communities in composting systems adapt to high nutrient loads, increasing decomposition efficiency. Introduced plants form new microbial or insect partnerships that influence establishment and spread. In vacant lots and other unmanaged sites, spontaneous communities form unexpected partnerships—examples of what some ecologists describe as emergent mutualisms.

A growing body of research is now exploring how symbioses evolve in cities. Key questions include how anthropogenic stress affects the stability and fitness outcomes of cooperative interactions, which traits enable persistence or transformation, and how these partnerships shape ecosystem function. Urban environments bring novel combinations of stress and opportunity that drive these dynamics.

Recognizing symbiosis as an evolutionary strategy reveals how urban systems shape both survival and interdependence. This perspective can inform the design of cities that cultivate adaptive, resilient forms of cooperation across species.

Multispecies Symbioses

In my class, our discussions about possible, futuristic symbiotic relationships in the city were revelatory. These animated conversations crystallized into four scenarios—distinct multispecies partnerships that could reshape life in the city.

Scenario 1: Plant–pollinator–microbe symbioses in urban parks. In nature, a complex relationship exists among plants, pollinators (such as bees), and the microbes inhabiting these pollinators. Microbes enhance plant nutrient absorption by aiding the metabolism of complex sugars in pollen while modulating pollinator behavior to optimize plant–pollinator interactions. However, in an urban park, this relationship faces new challenges. Antibiotic residues in urban water systems can destabilize microbial communities, reducing pollination efficiency. Urban gardens designed to attract native pollinators inadvertently simplify microbial communities through pesticides and fertilizers, disrupting traditional behavioral cues for efficient pollination. This raises questions: Can urban pressures trigger the rapid evolution of new microbial strains that forge novel relationships with their hosts? For example, might urban microbial strains evolve to degrade antibiotic residues, restoring pollination efficiency? Could these new strains enhance nutrient absorption and plant reproductive success in the urban context? What are the long-term impacts of such shifts on the resilience and biodiversity of urban ecosystems?

Scenario 2: Algae–coral–bacteria interactions in artificial reefs. Coral reefs are defined by a symbiotic relationship between the coral host, symbiotic algae, and microbial communities in their mucus layer, crucial for nutrient cycling, coral health, and reef stability. Artificial reefs in coastal cities aim to replicate these relationships, but reef engineers often overlook the complexity of microbial interactions. Polluted urban runoff can disrupt microbial community balance, making reefs more disease-prone. This raises questions: Could chronic pollutant exposure lead to the evolution of stress-resistant coral or algal phenotypes that can thrive in polluted environments? For instance, could these new phenotypes enhance nutrient cycling and coral health, increasing reef resilience? What effects would such evolution have on the larger marine ecosystem? Would it create more robust and adaptable coral communities?

Scenario 3: Urban soil–microbial community partnership. In natural soils, intricate relationships among bacteria, fungi, and plants support plant growth, nutrient cycling, and soil structure, affecting everything from plant health to soil fertility. Urban agriculture attempts to harness these symbiotic benefits but often fails due to soil degradation and the absence of native soil fauna. This leads to the disruption of the soil microbiome, affecting plant health and yield. Could urban conditions select for plants that are less reliant on complex soil communities but more dependent on specific urban-adapted microbes? For example, might we see the emergence of microbial communities that are particularly effective at breaking down urban pollutants, enhancing plant growth and resilience? Could these new partnerships improve the sustainability and productivity of urban agriculture?

Scenario 4: The urban microbiome. The human gut hosts a diverse range of bacteria, fungi, and viruses that interact with each other and the human body, influencing digestion, immune function, behavior, and cognitive functions. The urban lifestyle, characterized by diets high in processed foods and low in fiber, significantly impacts the gut microbiome. Additionally, air and water pollution introduce variables that disrupt this internal ecosystem. Could the pressures of urban living drive the evolution of a distinct "urban microbiome" adapted to the challenges and opportunities of city life? For instance, might we see the development of microbial strains that are more efficient at metabolizing processed foods or detoxifying pollutants, thereby enhancing urban health? How might such a shift impact public health and health care systems? Might it lead to new probiotic treatments and dietary recommendations tailored for urban populations?

Urban environments present unique challenges and opportunities for multispecies symbioses. As cities evolve, so will the eco-evolutionary dynamics within them, necessitating adaptive strategies attuned to these changing relationships. Human activities through traditional technologies have often disrupted natural processes, leading to the perception of urbanization as inherently destructive—an overly simplistic perspective. Urbanization can also be a catalyst for intricate networks of biotic interactions involving humans and other organisms. Recognizing this potential broadens our perspective beyond the negative effects of urbanization, highlighting the capacity of human activities to foster beneficial symbioses and contribute to maintaining ecosystem resilience and evolutionary potential.

The Symbiotic Niche

A species' realized niche—the range of environmental conditions and resources it actually occupies due to biotic interactions—may be affected by not only *competitive* interactions but also *mutualistic* interactions. Mutualistic interactions among species expand their niche space. For example, in the symbiotic relationship between mycorrhizal fungi and green plants, the fungus colonizes the root of a host plant, increasing water and nutrient absorption, while receiving carbohydrates produced by the plant via photosynthesis. This symbiosis allows the fungus to expand its range and the host plants to grow in lower-phosphorus soils and drier climates. Organisms must adapt to changes that they and their cohabitants cause in their environments through both competition and symbiosis. How do urban settlements change the mutualistic niche? And how does the emergent mutualistic niche alter symbiotic relationships among species?

The answers to these questions hinge on whether niche construction is viewed as a fundamental evolutionary process or merely as an outcome. Do selection pressures modified by environmental changes from human niche construction influence the evolutionary trajectories of populations? Niche construction is ubiquitous among living organisms and significantly shapes ecosystems. Evolutionary biologists debate whether habitat construction activities have evolutionary consequences comparable to those of natural selection. Proponents of niche construction theory argue that altering the environmental conditions that influence selection pressures changes the evolutionary dynamic, making niche construction an evolutionary process comparable to natural selection. Skeptics of this theory contend that niche construction's effects on evolutionary change are not unique but are generally a potential source of genetic covariance between phenotypic traits and fitness on which natural selection acts.[2]

All organisms modify their environments, with ecological and evolutionary consequences for other organisms.[3] They do so by affecting their fitness.[4] The controversy among evolutionary biologists is whether niche construction is a qualitatively different source of evolutionary change compared to other environmental sources.

To empirically settle this controversy, evolutionary biologist Blake Matthew and his team proposed that for niche construction to be a qualitatively different source of evolutionary consequences, it must satisfy three criteria: (1) An organism (a candidate niche constructor) must significantly modify environmental conditions, (2) these environmental modifications must influence selection pressures on a recipient of niche construction (i.e., an individual organism or population affected by the environmental changes), and (3) there must be a detectable evolutionary response in the recipient caused by the environmental modification of the niche constructor.[5]

The idea of niche construction as an evolutionary process can be traced back to Harvard evolutionary biologist Richard Lewontin, who, in the 1980s, described the ubiquity of niche construction in nature and the reciprocal interactions among genes, organisms, and environments.[6] Lewontin argued that "organisms do not *adapt* to their environments: they *construct* them."[7] Oxford biologist John Odling-Smee coined the term "niche construction" in 1988 and argued that niche construction is a fundamental evolutionary process.[8] In his 2003 book, *Niche Construction: The Neglected Process in Evolution*, Odling-Smee contends that organisms systematically modify selection pressures by constructing their niches, generating evolutionary feedback that can produce novel evolutionary outcomes.[9]

Along the same lines, Kevin Lala, Odling-Smee, and John Endler argue that organisms systematically and reliably change environments, imposing a consistent statistical bias on selection.[10] Analyzing more than 1,200 spatiotemporal datasets of selection in response to constructed versus nonconstructed sources, population geneticist Andrew Clark and colleagues found evidence that selective environments created by organisms differ from those arising from nonconstructed sources.[11] This complex interplay between niche construction and eco-evolutionary dynamics requires a novel view of evolution.

Symbiosis is both an evolutionary mechanism and an agent of niche construction. Symbiosis can generate evolutionary novelties by extending the phenotypic niche space of organisms. The symbiotic niche sets the stage for the emergence of new eco-evolutionary dynamics and feedbacks. The evolutionary consequences of the co-constructed niche can also alter symbiotic relationships among the co-constructor organisms and among other symbionts. Humans, in building their habitat, modify all such complex and delicate webs of reciprocal interactions among and between organisms, their partnerships, and their environment. Cities and their inhabitants are no exception to these reciprocal dynamics.

Urban Fungi's Lost Partner

Symbiotic relationships with mycorrhizal fungi are critical for plant health and establishment success, enabling plants to thrive in rapidly changing environments. However, mycorrhizae's ability to colonize plant roots is susceptible to microclimate changes, soil nutrient availability, and soil contamination. Despite the fungi's high resilience to disturbances, such as clear-cutting and weathering,[12] highly disturbed urban soils and increased deposition of nitrogen might affect the symbiotic relationship between the fungi and the plants.

Climate change can displace ectomycorrhizal fungi to other regions, leading to the loss of plant–fungi partnerships. The key factors affecting these symbioses are climate variables, such as temperature and moisture, which regulate organic matter decomposition and nutrient availability. Temperature and soil acidity similarly shape the interaction between plants and nitrogen-fixing bacteria that convert atmospheric nitrogen into forms usable for plant growth. Together, climate change and urbanization can shift plant species composition, disrupting existing fungal associations and redistributing symbiotic partners across landscapes. Forest land conversion to agriculture and cities, as well as climate change, have significantly challenged such a long-established partnership. These changes raise a broader set of questions: How does urbanization reorganize symbiotic interactions, and what new forms of symbiosis might emerge as species distributions, abundances, and resource constraints are reconfigured?

Finding New Partners

In cities, human influence creates a myriad of diverse and unique environmental conditions, fostering the emergence of novel ecological niches. These conditions offer opportunities for new partnerships among species, highlighting the inherent evolutionary capacity of life forms to adapt, innovate, and respond to change. Cities evolve through complex interactions among numerous agents, processes, and environments, leading to innumerable possible ecological outcomes and evolutionary changes.

Evolution is a crucial ecological process that alters species interactions and ultimately affects biodiversity and ecosystem dynamics.[13] Human-driven ecological and evolutionary changes in cities can directly affect symbiosis by displacing existing symbiont partners and indirectly by altering symbiont relationships. For example, removal of native vegetation and introduction of impervious surfaces can disrupt existing plant–mycorrhizal associations. As native plants disappear, the fungi that depend on them may also decline, leading to the loss of these beneficial partnerships. Conversely, nonnative plant species introduced to urban areas may bring their own symbiotic fungi, creating new relationships that alter the local soil microbiome.

Recognizing the potential for new symbiotic relationships in urban settings highlights the potential transformative role urban ecosystems can play in driving evolutionary change, enhancing biodiversity, and supporting more resilient urban environments.

Evolution of Urban Symbioses

Urban ecosystems exhibit unique heterogeneity due to the combination of natural and engineered landscape elements. This heterogeneity can foster greater niche differentiation and relatively high species diversity. Hybrid environments house diverse flora and fauna, enabling unexpected interactions, adaptations, and the emergence of novel ecological relationships that shape eco-evolutionary dynamics. Urban environments with their novel niches, newly introduced partners, and altered relationships present opportunities for symbiogenesis (the process by which new forms of life arise from symbiosis). Yet urbanization is most often interpreted as a source of ecological disruption, a view that tends to overshadow the potentially transformative role of urban eco-evolutionary dynamics.

Symbioses are complex, coevolved relationships that arise from the interplay and coordination of diverse genetic blueprints and play pivotal roles in ecosystem function.[14] They contribute significantly to the creation of biological novelty, the shaping of biodiversity, and the facilitation of key transitions within Earth's ecosystem. These relationships are not only foundational to natural processes, such as biogeochemical cycles, but also have tangible implications for agriculture, soil integrity, forest health, and human well-being. Evolutionary biologist Erik Hom and complexity scientist Alexandra Penn characterize symbiosis as "shared genetic fate of two or more organisms via physical association"[15]—a more refined understanding beyond traditional labels like "mutualism," "commensalism," or "parasitism." They assert that symbiotic relationships are fluid, multifaceted, and heavily influenced by environmental contexts, underscoring the need to examine the level of integration and homeostatic processes in discerning how adaptive coevolution impacts symbioses.

New symbioses in urban environments can emerge via multiple mechanisms. Urban heat islands, green spaces, and waste sites constitute novel niches with distinct climatic, environmental, and structural conditions that may harbor unique symbiotic pairings not present elsewhere.[16] For instance, artificial structures such as buildings and bridges create spaces and modify sunlight, temperature, and moisture levels, fostering conditions in which novel symbiotic interactions among microbial communities, plants, and animals can emerge.[17] Urban areas, as centers of trade and transportation, can enable mixing of nonnative species, thus initiating new symbiotic relationships.[18] The urban environment, dense in both human population and biodiversity, may catalyze new partnerships between potential symbiotic partners, a phenomenon reflected in the richness of urban green spaces.[19]

Urbanization leads to shifts in selection pressure, which may favor symbiosis. Intense competition for resources may promote beneficial partnerships, enhancing survival and reproduction.[20] Pollution, a characteristic urban stressor, may foster detoxifying symbioses—symbiont-mediated modification of a toxin that benefits the host.[21]

Rather than seeing urbanization as a disruptor of natural symbioses, we can envision symbiosis as an agent of ecosystem change, with humans emerging as new partners. Hom and Penn offer a unique perspective on symbiosis in the Anthropocene.[22] Using farming as an example, they suggest that farming practices throughout history have led to the formation of new, persistent relationships among humans, crops, microbes, invertebrates, and livestock, leading to a coevolving web of interactions within the agricultural ecosystem. While intensive agricultural practices have disrupted the symbiotic relationship between rhizobia and crop plants, an alternative view could suggest that humans, through the use of artificial fertilizers, have "effectively" replaced rhizobia as a nitrogen source for crops, implying that farming could be viewed as a new form of symbiosis between humans and cultivated species.

Under rapid environmental change, species may become obligate mutualists, indicating that their mutual dependency is essential for survival. A study involving the yeast *Saccharomyces cerevisiae* and green alga *Chlamydomonas reinhardtii* demonstrated that obligate mutualism can spontaneously arise in an environment where both carbon and nitrogen exchange is necessary.[23] This finding shows that environmental changes can lead free-living species to form new symbiotic partnerships without requiring prior coevolutionary adaptation. The findings provide crucial insights into the process of symbiosis and offer urban evolutionary biologists a basis for designing experiments to study how environmental shifts in urban environments might foster novel mutualistic relationships and symbiosis.

Building on this perspective, human technological interventions present an unprecedented opportunity to foster novel symbiotic relationships, enhancing ecosystems' adaptive and evolutionary potential. Among the emerging approaches that seek to work with ecological processes in urban contexts, nature-based solutions offer a particularly promising pathway. When designed to facilitate symbiosis and account for eco-evolutionary dynamics, interventions such as green roofs and urban wetlands can create habitats that support diverse species while contributing to biodiversity and human well-being. Urbanization, often perceived as disruptive to native ecosystems, can instead serve as a platform for hybrid networks of interaction between humans and other organisms. The challenge lies in deliberately channeling human agency toward fostering these connections, so that human needs are met in ways that sustain the health and resilience of the biosphere.

The direct experience of accelerating climate change has become a powerful catalyst for behavioral adaptation in humans, underscoring the profound interconnectedness between human activity and ecological systems. Witnessing extreme events—record-breaking heatwaves, intensifying storms, and rising sea levels—has sparked a collective realization of the urgent need to rethink our relationship with the natural world. This awareness is deeply rooted in lived experiences, such as coping with floods and disrupted food supplies, which expose the fragility and interdependence of urban ecosystems. These experiences are reshaping behavior, inspiring shifts toward sustainable practices such as reducing energy consumption, adopting renewable energy, and engaging in collective climate action.

Concrete examples of this potential abound. Green infrastructure mitigates urban heat island effects and improves stormwater management while providing habitats for birds, insects, and plants, fostering new symbiotic relationships and enriching urban biodiversity. In marine environments, artificial reefs constructed from discarded materials, such as concrete or decommissioned ships, create habitats that support diverse marine life, establishing new ecological communities and facilitating symbioses among marine species. Similarly, in agriculture, the application of biofertilizers—microorganisms that enhance plant nutrient absorption—represents a deliberate attempt to forge beneficial partnerships. These microbial inoculants not only increase crop productivity but also improve soil health, nurturing a mutually advantageous relationship among crops, microorganisms, and the soil ecosystem.

The convergence of technological innovation and shifting human behavior can further foster symbioses between humans and other organisms. For example, the integration of smart technologies in urban agriculture enables resource efficiency while fostering ecological balance. In cities like Singapore, urban farms equipped with precision agriculture technologies—such as IoT (Internet of Things) sensors and automated irrigation systems—optimize water and nutrient use while creating habitats for pollinators like bees and butterflies. Vertical farming systems integrated into urban landscapes reduce the spatial footprint of food production while enhancing air quality and mitigating the urban heat island effect. These interventions highlight the capacity of urban spaces to serve as dynamic arenas for the emergence of symbiotic relationships that foster resilience.

Symbiotic relationships, while predominantly characterized by mutual benefit, inherently involve conflicts and tensions that require negotiation and balance. This dynamic is akin to the "holobiont" concept, which refers to a host organism and its microbial symbionts achieving a state of homeostasis through intricate interactions. Urban ecosystems mirror this complexity. Diverse stakeholders—residents, businesses, local authorities, and nonprofits—coexist within shared spaces, each with distinct goals and responsibilities. Like the components of a holobiont, these entities must navigate competing interests to achieve a cohesive, functioning urban system.

For instance, local authorities might reconcile economic development with the community's demand for green spaces, while businesses and nonprofits collaborate to address shared goals like sustainability and social equity. This negotiation process fosters a form of urban homeostasis, where balance and functionality are maintained amid divergent priorities. Recognizing and nurturing these dynamics enables cities to evolve as resilient, symbiotic systems capable of addressing contemporary challenges while fostering innovative ecological partnerships.

By embracing these principles, we can reimagine urban spaces as hubs for symbiotic innovation, where human ingenuity aligns with ecological resilience to create thriving, interconnected communities. This vision not only can addresses the immediate impacts of climate change but also lays the foundation for a sustainable, adaptive future where humans and nature coevolve in harmony.

4 Evolutionary Resilience

During the question-and-answer session at the end of my talk, one question stood out: "If urban species are evolving rapidly in today's changing environments, then how do we build cities that are resilient?" Over the previous year, I had spoken to city planners—both scholars and practitioners—across the world, and I had one central message: We must rethink resilience from an eco-evolutionary perspective and embrace a dynamic view of the biosphere.

Urban resilience is the inherent property of a social-ecological system that enables it to maintain its essential functions in the face of dynamic change.[1] An eco-evolutionary perspective reshapes our understanding of how cities interact with ecosystems to sustain their adaptive capacity. Eco-evolutionary resilience—defined as the capacity of ecosystems to maintain ecological function through eco-evolutionary processes under changing environmental conditions—offers a more robust foundation for understanding the long-term adaptability and stability of urban social-ecological systems.[2] It challenges the static mindset that still dominates urban planning, urging us to shift from trying to keep nature still, just as things are, toward strategies that enable evolutionary change, equipping cities to navigate uncertainty and expand their capacity to adapt.

I answered: "Resilience depends on preserving evolutionary potential. Evolution is the mechanism through which the natural world adapts to change. It underpins the ability to respond to shifting conditions. The greater the genetic diversity within a population, the greater the likelihood that some individuals will possess traits suited to shifting habitat conditions."

To my delight, the notion of linking the evolving traits of urban lizards, fish, insects, and plants to city planning has sparked curiosity, not skepticism. Despite the unequivocal differences between the Global South and the Global North in pace of urbanization and the stage of development, all city planners are faced with the same challenge: how to adapt to rapid environmental change.

Bringing evolution into view changes everything. The reference condition becomes a moving target. It expands the number of factors to plan for in conservation and adaptation because of the multiple interactions that drive natural selection and affect eco-evolutionary outcomes. It amplifies the variability and uncertainty that must be taken into consideration when making decisions because the outcomes are highly unpredictable. And, especially, it changes how we understand the role and strategies of city planning, expanding its drawing board. Because of the complex nature of urban ecological and evolutionary dynamics, the rules governing selection and adaptation are redefined. Humans become protagonists in a new eco-evolutionary play.

In ecology, resilience is the capacity of a system to absorb disturbance from change and reorganize while maintaining essentially the same function, structure, identity, and feedback.[3] For cities, as for ecosystems, their complexity and diversity provide the foundations to adapt to change. Species play an important role in regulating ecosystem function—the physical, chemical, and biological processes that control the fluxes of energy, nutrients, and organic matter that maintain Earth's life support systems and provide essential services for human well-being. Observed changes in species traits might affect ecosystem function by altering population dynamics, community structure, and species interactions, thereby influencing evolution and creating new eco-evolutionary feedbacks. This is because an organism's functional traits—morphological, physiological, phenological, and behavioral—influence its fitness and enable it to adapt to environmental changes.

In building resilient cities, we must draw on our understanding of how evolution provides robustness and resilience to biological systems.[4] Evolution enables populations to withstand environmental changes by selecting for genotypes that best fit novel environments, conferring a fitness advantage. Genetic variation underlying a phenotypic trait that confers a survival or reproductive advantage is more likely to be passed on to the next generation. Life has evolved over a long time toward complexity and resilience, strengthening the feedback mechanisms that stabilize the climate system and biogeochemical cycles.[5] Yet evolution does not always imply that more complexity is added to life. Both simpler and more complex forms of life may emerge from evolution. Parasites, such as tapeworms, lost their digestive tract, brain, and many sensory organs that are not needed for living inside the intestines of other animals. Many aquatic animals (whales and dolphins) and terrestrial animals (snakes and lizards) have significantly reduced or lost their limbs because the limbs were redundant in their new habitats. Evolution of more complex life occurs when it's necessary for survival and reproduction.

Urbanization challenges the fate of populations and simultaneously accelerates their evolution at a pace visible within our lifetimes. Rapid evolution in species traits might have significant effects on the ecosystem functions that sustain human well-being, such as nutrient cycling, pollination, carbon sequestration, water and air purification, and food production. And such changes will affect the resilience of urban communities and ecosystems. Most ecosystem functions and the services they provide to urban communities involve species interactions across trophic levels, for example, nutrient cycling, pollination, climate regulation, and pest control. Urbanization also exerts selective pressures on traits that underlie species interactions (e.g., foraging traits and defense traits), driving changes in community dynamics that control ecosystem functions and causing cascading evolutionary and ecological changes across trophic levels, which might reduce community stability and resilience.

Urbanization, while often seen as a challenge, can also create opportunities for resilience on a coevolving, hybrid planet. The capacity of ecological community restructuring to enhance urban ecosystem resilience depends on how urban development shapes the strength of interactions between evolutionary processes and community dynamics—specifically, the degree to which species' interactions affect their respective survival and reproduction—underpinning essential ecosystem functions. For instance, shifts in community structure due to evolved dispersal traits can significantly affect pollinator dynamics, plant reproduction, and local food production. On the other hand, constructed networks can extend the reach of seed dispersal across greater distances. Restoring and creating corridors between natural habitats can enhance the movement of pollinators and other species, improving their capacity to disperse pollen and seeds. The strength of trophic cascades, which can be pivotal in determining how local ecosystems respond to environmental changes, affects their ability to maintain stability and regulate processes such rainwater capture, thermal buffering, and vector-borne disease dynamics. Small perturbations at one trophic level can reverberate through the food chain, causing extensive impacts. Experimental evidence shows that evolution can shift ecological tipping points, altering the thresholds at which communities persist or collapse.[6] Understanding the role of urbanization in shaping these trophic cascades is key to maintaining eco-evolutionary resilience.

Evolutionary Fitness in Cities

Humans, as part of the animal kingdom, rely on primary producers—plants and algae—that, through photosynthesis, convert sunlight, carbon dioxide, and water into food. The evolution of plant traits that govern the uptake and use of carbon, water, and nutrients—such as plant metabolism, leaf structure, and nutrient allocation—directly impacts plant growth and, consequently, plant fitness. Recent studies show that plants in urban environments exhibit distinct shifts in leaf functional traits.[7] Cities, with their unique pressures, tend to select for plant traits that improve resource acquisition, enabling survival in harsh conditions. For example, urban plants often exhibit lower stomatal density, reduced leaf width, and decreased phosphorus concentration compared to their rural counterparts—adaptations likely driven by elevated temperatures, pollution, and altered soil nutrient content. These changes in urban plant leaves, such as smaller leaf size, restrict carbon dioxide exchange[8] while enhancing water-use efficiency by minimizing water loss and regulating heat.[9]

However, this trend is not uniform across a city; it varies with the sensitivity and age of different plants, as well as with land-use types and the intensity of human activities. Variability in physiological traits that govern photosynthesis and carbon uptake, such as chlorophyll content or photosynthetic rate, is often strongly influenced by the developmental stage of the leaf—with performance typically increasing as leaves mature, peaking, and then declining with senescence.[10] In contrast, variability in leaf morphological traits—such as size, shape, and structure—is more strongly shaped by environmental context, including land use and microclimatic conditions.[11] These morphological traits reflect long-term adaptations that enable plants to optimize resource acquisition across diverse environments. In response to environmental pressures, plants employ different survival strategies—prioritizing growth, reproduction, or defense—depending on the specific challenges encountered.

A study by Yuebo Su and colleagues on the Chinese red pine (*Pinus tabuliformis*), a native pine of northern China, along an urban–rural gradient in

Beijing, found that pines growing near roads had smaller leaves, lower stomatal density, and reduced leaf area compared to those growing in parks.[12] From an economic perspective, plants exposed to air pollution adopt a low-risk investment strategy. Urban roads and their associated pollutants significantly impact the performance of plant leaves, affecting growth, photosynthesis, and essential ecosystem functions such as carbon sequestration and air purification. Thus, the functional traits of plants offer insights into how they make trade-offs in response to the diverse stressors characteristic of urban landscapes, providing crucial information for urban tree planting.

As humans continuously adjust their resource and energy allocation in response to changing conditions, so do other organisms. In resource-limited environments and under multiple stressors, plants optimize resource allocation among functional traits. The evergreen shrub *Euonymus japonicus*, a common species in Beijing's parks and gardens, exhibits a trade-off strategy between defense and growth. Its reduced leaf area suggests that the plant allocates most of its resources to minimizing damage from atmospheric pollution.[13] Known for its ability to thrive in the dry and cold conditions typical of northern cities like Beijing, *Euonymus japonicus* is also adept at intercepting and capturing particulate matter, a major urban air pollutant caused by vehicles.

Animals, too, face trade-offs in urban environments. An animal that invests more energy in growth and development may have less energy available for reproduction and vice versa. Prey species may trade investment in defensive traits for energy allocated to other functions. Similarly, predators might trade improved problem-solving abilities for reduced physical capabilities, such as speed or strength, or balance increased aggression with reduced sociality. Urban predators must often weigh the benefits of defending their territory and resources against the risks of being more aggressive, including potential injury or conflict with humans. Another trade-off occurs when a shift in diet toward a broader range of foods leads to compromises in nutritional quality and potentially larger body size.

Intraspecific trait changes—shifts in behavior or morphology within a species due to urbanization—can drive variations in species interactions and population dynamics, which, in turn, can feed back to influence further trait changes in populations. Trait variation resulting from selection pressures in urban environments may depend on predator density, resource availability, and habitat conditions. This feedback loop between abundance and trait dynamics is particularly pronounced in urban settings, where the increased co-occurrence of wildlife species compresses the space and time shared by communities, intensifying species interactions.

A crucial question is whether rapid evolution driven by urbanization will determine the persistence or extinction of a population. How will the environmental challenges faced by organisms in cities affect their resilience and the resilience of the communities that depend on them? While the evolution of species can often prevent extinction, it is not guaranteed. Adaptation does not necessarily reduce extinction risks across all taxa in a region. The outcome depends on the pace of urban development, the concurrent rate of climate change, and the organisms' ability to adapt. Natural selection promotes the evolution of traits that help a population thrive in its local environment, leading to higher fitness, population growth, and persistence. However, it is the rate of evolutionary fitness that determines whether adaptation can outpace maladaptation.[14]

The compression of time and space for species interactions in urban environments may accelerate or decelerate evolutionary change by altering the strength of interspecies interactions—such as competition, predation, and mutualism—that drive natural selection on fitness-related traits. In urban settings, species often encounter each other more frequently and in closer proximity than in natural environments, intensifying opportunities for interaction. Between two species competing for the same resources, the more efficient at acquiring and using them will have a survival and reproductive advantage. This likely leads to the evolution of traits that allow one species to outcompete the other. Similarly, prey species may evolve traits that enhance their ability to evade or defend against predators.

From an ecological standpoint, urbanization reduces top-down control while increasing the influence of bottom-up mechanisms through the greater availability of anthropogenic resources (e.g., food, microclimate) provided by urban environments.[15] A shift from top-down to bottom-up processes can feed back on population and community demographics, either relaxing or strengthening selection pressures, which may reinforce eco-evolutionary feedback by generating increasing directional selection across multiple trophic levels. Alternatively, urbanization can dampen selection strength by making it easier for individuals to survive and reproduce (e.g., through resource provision), thereby reducing selective pressures and altering the environment such that certain traits are no longer advantageous—ultimately decreasing the opportunity for selection.[16]

Understanding the relationships between contemporary evolution and population fitness is critical for predicting how populations will respond to urbanization and environmental change and for informing conservation strategies to protect biodiversity and maintain ecosystem function. However, evolutionary responses that increase fitness or support population persistence will not necessarily sustain the ecological functions on which resilience depends. Eco-evolutionary resilience therefore depends not simply on adaptation, but on whether adaptive change sustain the ecological dynamics that maintain function.

Linking evolution to ecosystem resilience is far from simple. Resilience to environmental change depends on ecological and evolutionary processes operating across all levels of biological organization and scales. Urban environments influence species traits through multiple selection pressures, altering species interactions across local, regional, and global scales.[17] Increased temperatures, pollution, and artificial light (to name a few selection pressures) create new challenges for both animals and plants, directly impacting how species traits respond to these new environments. Under the multiple pressures they face in urban environments, organisms must decide which functional traits to prioritize and how to allocate resources to balance survival, development, and reproduction.

While natural selection typically favors traits that increase fitness in current local conditions, this may not always confer long-term resilience. In rapidly changing environments—especially under accelerating climate change—species with traits preadapted to future conditions may have a distinct advantage over those optimized for the present. For example, genotypes with greater tolerance to heat, drought, or pollutants may outcompete locally adapted populations as environmental thresholds are crossed. This raises important questions for conservation and urban design: Should we prioritize preserving locally adapted populations or facilitate the migration and establishment of genotypes better suited to emerging conditions? Understanding which traits enhance future fitness—and how these traits evolve—will be critical for maintaining species persistence and ecosystem function in cities.

Foresight and Evolutionary Strategies

Imagine navigating a maze, knowing that every turn you take isn't just about your immediate survival but could shape the evolutionary path of future generations. This ability to anticipate and prepare for what lies ahead—what some call "mental time travel"—might seem like a uniquely human trait. But what if foresight isn't unique to humans? What if the capacity to anticipate and plan for the future has evolved across multiple species—and plays a vital role in their survival and adaptation?

Life history theory posits that organisms evolve traits to balance survival and reproduction across their lifespan. These trade-offs shape evolutionary fitness. But how do organisms weigh future risks in a changing world—and could foresight itself shape these life history strategies?

This question challenges us to rethink what we know about evolution, particularly in light of how foresight—whether in humans or other species—can modify the outcomes of evolutionary processes. As we explore the role of foresight in both human and nonhuman life histories, we find ourselves on the brink of understanding a deeper connection between cognition and evolution, where the ability to envision the future might just be as crucial as the genetic blueprints passed down through generations.

As other organisms do, humans make trade-offs among different functional domains. Despite challenges in measuring and finding evidence of trade-offs in human life history, life history theory remains a dominant framework in evolutionary anthropology and demography for explaining fertility and mortality patterns in humans and accounting for major shifts during demographic transitions such as reduction in fertility. Yet there is evidence that, in humans, other factors (such as variable access to resources among populations, variable cultural and economic influences, and social norms) may play key roles. Furthermore, the emergence and rapid evolution of technology has shaped human capacity, and it is one major explanation for the demographic transition of human societies.

What is less understood is how *foresight* has evolved in humans and how it shapes intergenerational trade-offs in life history strategies. And, is "mental time travel" a cognitive ability unique to the human species? Some scientists believe so. Psychologists Thomas Suddendorf and Michael Corballis define foresight as a human's ability to insert the past or future recursively into their thinking.[18] They see such ability as a discontinuity between ourselves and other animals; "thinking about thinking" might be a quality that distinguishes our species from all other creatures on the planet. Yet growing evidence shows that chimpanzees, corvids, and even insects also anticipate future outcomes of their actions and build tools to solve problems. Foresight appears to be an evolutionary advantage that allows many species to thrive.

Early primate studies revealed that chimpanzees save food and tools for future use. More recently, similar experiments conducted on crows and ravens found that they also possess these skills as a problem-solving strategy. Scientists have observed that many insects, such as grasshoppers and dragonflies, use internal models of the world to guide their actions and predict the outcome, and scientists speculate that "forward-planning" neural circuits, rather than preprogrammed routines, might explain this behavioral flexibility.[19] Recent evidence that rats can intentionally engage hippocampal representations for navigation and object manipulation in a virtual space supports the hypothesis that rodents may possess imaginative capacity.[20]

Honeybees and spiders build their nests by planning ahead to achieve their goals. Similarly, ants show great flexibility in choosing types of objects and tools and selecting the most optimal to achieve a desired target. The evidence of these complex behaviors has accumulated over many years, but only recently have scientists started to explore the possibility that foresight is at play.

More than 200 years ago, Swiss entomologist Francois Huber showed, through a series of experiments in which he challenged bees with various obstacles to their hive-building process, that honeybees have the capacity to employ some sort of master plan to proactively tackle challenges.[21] Huber placed glass panels into the path of honeybees building a hive. The bees' response was quite extraordinary. They adjusted their construction plan to strategically avoid the unsuitable, slippery surface. Recent studies provide evidence that bees' complex behavioral flexibility is the result of their capacity to identify the target of their actions and choose the most efficient strategy among multiple envisioned strategies. Although we do not yet have direct evidence of arthropods' neural circuitry that would support their imagination of possible futures, neurobiologists Clint Perry and Lars Chittka hypothesize that foresight is the most plausible explanation of arthropods' ability to solve problems never experienced throughout their species' evolutionary history.[22]

If the ability to think ahead is not unique to humans, an intriguing question is how human culture—and the ability to acquire knowledge and skills from others—amplifies foresight and thereby enables humans' unmatched capacity for problem-solving. Humans build iteratively on that reservoir of pooled knowledge over long periods, which gives our species an edge.

Culture seems to separate us from the rest of nature, and yet that culture, too, must be a product of evolution. The scope and scale of human reach and the ability to deliberate have enabled humans to develop increasingly sophisticated technologies that have allowed human societies to adapt to nearly every habitat on Earth.[23] The role that cities played in the evolution of culture and the emergence of complex society is still debated. What we know is that the emergence of the first cities in ancient Mesopotamia is inextricably linked to the evolving interactions between people and the environment, which have both enabled city building and transformed human society.[24]

Whether urbanization has driven the emergence of human complex society or arisen from it, the structural similarities among urban systems across regions, cultures, periods, and circumstances point to a remarkable convergence in the ways that humans have adapted to diverse environments. However, this convergence may lead us to view urbanization as a homogeneous phenomenon and to underestimate the diversity of its eco-evolutionary dynamics. Closer examination reveals that Earth's future is shaped by multiple interacting forces, which could generate divergent trajectories rather than a single, predetermined path.

Cities can each be viewed as a microcosm of a possible future Earth. The ways in which urban eco-evolutionary dynamics transform biodiversity and ecosystem productivity illustrate how the diversity we observe within and across cities might lead to different future outcomes. While urbanization is typically associated with loss of biodiversity, an increasing number of studies examining species and genetic diversity in urban areas challenges this expectations revealing idiosyncratic ecological responses[25] and evolutionary outcomes.[26] These findings prompt intriguing questions about what drives such inconsistencies and suggest that variable patterns of biodiversity may reflect differences in the emergent properties of urban landscapes.[27]

Urbanization affects biodiversity by altering processes, landforms, species, and sources of ecological variation through multiple mechanisms.[28] Yet these mechanisms and their effects are not uniform across landscapes and urban regions. The impact of urbanization on net primary productivity, the rate at which primary producers such as plants and algae generate biomass, can vary significantly across urban agglomerations depending on patterns of urban development. Alternative urban trajectories can make ecosystems less or more resilient by limiting or enabling different species to perform different functions and fill different niches.

The Paradox of Ecological Design

If urban environments fundamentally alter, on a contemporary time scale, the evolution of species and their adaptive capacity, how do we design and build cities that enable species to adapt and maintain resilience in the face of unpredictable eco-evolutionary futures? In urban planning, integrating population and community responses to anthropogenic change is no longer optional—it is essential. The challenge lies in designing resilient infrastructures that account for the complex, urban eco-evolutionary dynamics. How can we rethink ecological design to account for the inherent dynamics of coupled human–natural systems?

Despite the well-established understanding that ecological systems rarely maintain stable equilibria,[29] the assumption of environmental stationarity continues to pervade ecological design. This assumption is tied to the outdated belief that biological responses to changing environments remain stable over spatial and temporal scales.[30] Over 50 years ago, Canadian ecologist C. S. "Buzz" Holling articulated that ecosystems are in constant flux, with stability that is patchy and scale-dependent. "Change," Holling noted, "is neither constant and gradual nor consistently chaotic. Rather, it is episodic, with periods of slow accumulation punctuated by sudden reorganization. These events can shape trajectories far into the future."[31] Although several scholars have since adopted the view of ecosystems as open, multiple-equilibrium systems that are, to some extent, unpredictable, the prevailing design paradigm is still based on the assumption that stability and control are achievable—and even desirable—goals.

The primary objective of ecological design has traditionally been to achieve a stable, known condition. Even when designers acknowledge the inherent instability of natural systems and focus on identifying disturbances to minimize their impact, they often still operate under the assumption of a stable reference condition. This approach aims to build adaptive capacity to maintain a target state, but from an eco-evolutionary perspective, this is a paradox. Designing for a fixed state within a system that is inherently transient can reduce ecological resilience and lead to collapse. As Holling cautioned, "It is at least conceivable that the effective and responsible effort to provide a maximum sustained yield from a fish population or a nonfluctuating supply of water from a watershed might paradoxically increase the chance for extinctions."[32]

The underlying assumption of much ecological design is that ecosystem dynamics are relatively well understood, thresholds are constant, and outcomes are largely predictable. However, as Holling pointed out, knowledge of ecosystem dynamics is far from complete, and surprise is inevitable. Ecosystems are moving targets with multiple equilibria, meaning any attempt to control their dynamics is destined to fall short. Holling identified the root of the divide between ecology and design in their differing views of ecosystem dynamics and resilience. In his groundbreaking 1973 paper "Resilience and Stability of Ecological Systems," Holling defined ecological resilience as the "ability of a system to absorb changes and still persist," or "the amount of disturbance a system [can] take before it [shifts] into an alternative configuration."[33] In contrast, designers have often embraced the near-equilibrium concept of engineering resilience, which is rooted in deductive mathematical theory. This approach focuses on a single equilibrium and refers to the ability of a system to return to that state after being temporarily perturbed.

Despite mounting evidence that urban environments drive rapid evolutionary change—with direct consequences for ecosystem functions that underpin human well-being—urban design often treats biodiversity and ecosystem function as fixed endpoints. Urban wetlands, bioretention systems, rain gardens, and stormwater ponds are usually conceived to preserve existing biodiversity or re-create preurban conditions, assuming that species traits and interactions remain constant. These interventions aim to regulate stormwater, reduce heat, filter pollutants, and enhance human well-being, yet their long-term performance is shaped by eco-evolutionary dynamics rarely considered in their design. By generating novel habitats, they can restructure ecological networks, alter species interactions, and trigger evolutionary shifts in traits—changes that may strengthen or undermine the functions these systems are intended to support.

Ecological design assumes that organisms and interactions that underpin ecosystem function remain relatively stable over time. But evolution can alter the populations and traits that sustain these functions.[34] Pests may evolve resistance to introduced predators, undermining biocontrol.[35] Species planted for flood or erosion control may shift biomass allocation toward rapid aboveground growth at the expense of roots, reducing stability.[36] Evolution in photosynthetic capacity, growth, or defense traits can alter trees' ability to sequester carbon and mitigate air pollution.[37] Ignoring these dynamics risks undesirable evolutionary responses, destabilizing trade-offs, and disrupted interactions that erode long-term ecosystem performance.[38]

Many examples of observed urban-driven changes in organisms affect their ability to perform important ecosystem functions (figure 4.1).[39] In urban trees, shifts in drought tolerance, phenology, and pest-resistance trade-offs can influence carbon sequestration, urban cooling, and air quality improvement, shaping how cities respond to heat and pollution.[40] In bumblebees, increases in body size driven by urbanization can influence pollination efficiency with consequences for plant reproduction, crop productivity, and biodiversity.[41] Rapid evolution in root traits, salt tolerance, and belowground biomass in chairmaker's bulrush (*Schoenoplectus americanus*) modifies carbon accumulation and soil surface accretion, which are critical for the resilience of marsh ecosystems to rising sea levels.[42] In *Daphnia* spp., adaptations to urban heat islands, toxic cyanobacteria, and pesticide exposure alter physiological and life-history traits,[43] potentially influencing their role in water filtration and the regulation of algal blooms.[44] Meanwhile, climate-driven changes in oysters, including thermal tolerance, growth, shell integrity, and disease susceptibility, may alter shoreline protection, water filtration, and biodiversity support under ocean warming and sea-level rise.[45] Changes in microbial communities, such as antibiotic resistance and horizontal gene transfer, may also affect wastewater treatment, degradation of pollutants, and pathogen reduction in urban systems.[46]

From Biomimicry to Bio-Cooperation

If ecosystems are moving targets and knowledge is incomplete, how can designers pursue the goal of aligning the built environment with ecological dynamics? We may not be able to make a major shift in our thinking about cities unless we see ourselves as key players in evolution. Humans are active agents in niche construction, changing our environment in significant ways, and we actively participate in ecosystem functioning and evolutionary change. In cities, humans change everything: composition, processes, dynamics, functions. By building structures (skyscrapers, stadiums, factories) and infrastructure (underground plumbing and subways, overhead electrical wires, airports), humans redistribute organisms and change energy cycles and material cycles, leading to distinct and novel urban biogeochemistry and patterns of biodiversity. Like other species, humans engage in all forms of biotic interactions that play pivotal roles in the structure and functioning of ecological communities.

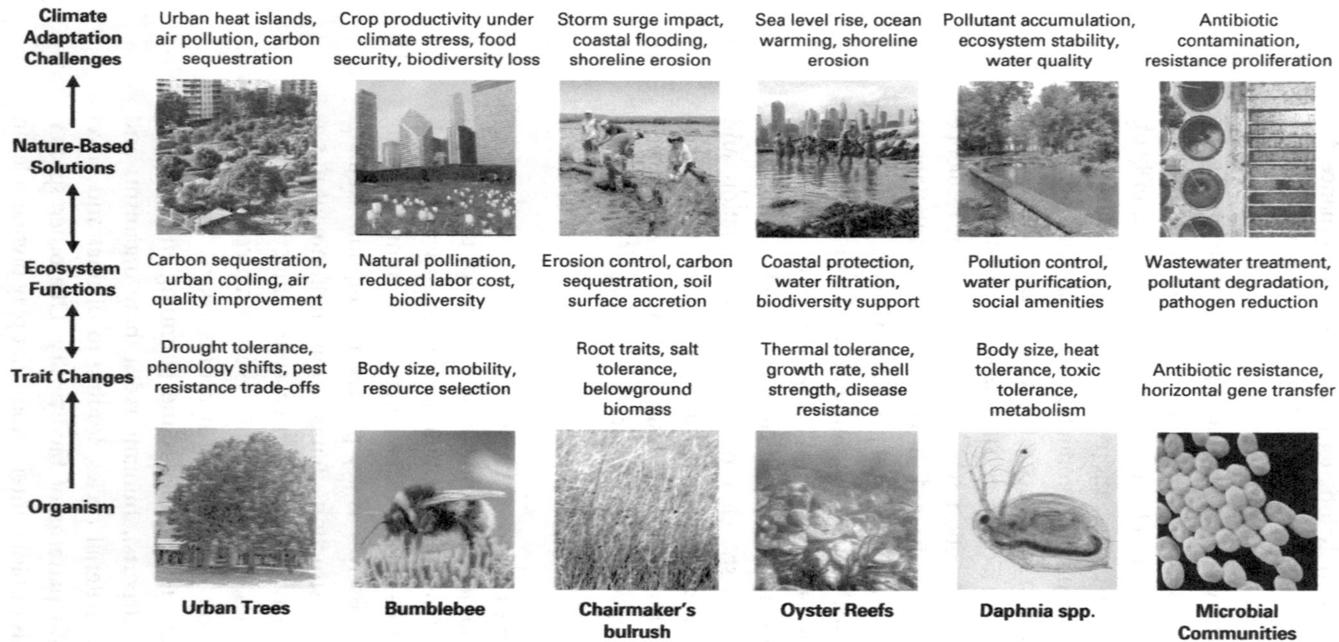

Figure 4.1

Examples of urban-driven evolutionary trait changes that alter organisms' ability to perform key ecosystem functions. *Source*: Author-created image, adapted from Alberti et al., *Science*, 2026.[47] Photographs: Urban Trees, plaza with urban trees (aerial view), Unsplash, Nerea Martí Sesarino. London plane tree (*Platanus × acerifolia*), iNaturalist, Helen Frost, public domain (CC0). Bumblebee, Millennium Park, Flickr, Mike Linksvayer. Bumblebee (*Bombus* spp.), Pixabay, Myriams-Fotos, public domain (CC0). Tidal marsh restoration at Edwin B. Forsythe National Wildlife Refuge, Flickr, Don Freiday/USFWS, public domain (PD). Chairmaker's bulrush, (*Schoenoplectus* spp.), Wikimedia, Gordon Leppig & Andrea J. Pickart, public domain. Oyster reef (*Crassostrea* spp.), Billion Oyster Project, © Steven DeWitt/Witness Tree Media. Oyster reef (*Crassostrea* spp.), Weekapaug Breachway, Wikimedia, NRCS Rhode Island, public domain (CC0). Daphnia (*Daphnia* spp.), public park pond and footbridge, Memorial Museum Passchendaele 1917, Zonnebeke, Belgium, Flickr, Edk7, (reproduced with permission). Daphnia (*Daphnia pulex*), iNaturalist, David Bird, public domain (CC0). Wastewater treatment plant (aerial view), Unsplash, Ivana Bandura. *Acinetobacter* sp., CDC/PHIL, public domain (PD).

Humans play key roles as dominant predators and consumers and as agents of physical disturbance. Yet the predominant view in ecology and earth science still places humans *outside* the biosphere. Rethinking eco-evolutionary dynamics in the presence of humans implies expanding the set of variables and conditions that cause eco-evolutionary change and the emergence of evolutionary and ecological innovations. Including humans in eco-evolutionary dynamics requires understanding the role of niche construction and cultural coevolution (through inheritance and social learning).

Ironically, along the same lines, the social sciences and humanities place nature and other organisms *outside* the human sphere. Biomimicry, an emerging design approach in which solutions emulate natural processes, is emblematic of such dichotomy. By mimicking nature, we place ourselves outside of nature. We mimic nature to reconnect with nature, by interpreting and replicating what nature does. But this view of nature excludes us. Such a view is so embedded in our vision, language, and perception that we speak about "protecting nature," "maintaining nature's provisions and ecosystem services to humanity," saving Earth "from us and for us," and imagining "an Earth without us."

Toward Urban Eco-Evolutionary Resilience

An eco-evolutionary perspective fundamentally transforms how we approach and practice city planning. Understanding how cities select for species traits and untangling the mechanisms that enable organisms to survive and reproduce will provide new insights for designers and planners to simultaneously mitigate the impacts of urbanization and climate change by expanding cities' adaptive capacity in ways that consider all affected organisms. An evolutionary perspective will help us see how historical dynamics have shaped the system's capacity for adaptation and reveal organisms' evolutionary potential. Evolution will affect how organisms respond to urbanization and climate change, altering intraspecific (within species) and interspecific (among species) interactions, as well as the ecosystem functions on which urban resilience depends.

Such a perspective shifts the focus of planning toward human–nature interactions and new cooperation between humans and the biosphere. Understanding cities from an eco-evolutionary perspective expands the dimensional space for human–nature relations and the potential pathways of their coexistence. Planning will then rely on principles that expand heterogeneity and support modularity (as opposed to generalized connectivity) and self-organization. By implementing eco-evolutionary principles, urban structures and infrastructures will support evolutionary potential.

Plants and animals make trade-offs based on the best information available to them: accessible resources, potential threats, and their ability to cope with threats. We learn about their adaptation and resilience by observing and measuring outcomes, the conditions under which they survive, and how their life cycles change. As we begin to account for their evolution, what we observe and measure changes with urbanization. And what makes learning about urban evolution more interesting is that *we* are a key agent in the complex, evolving, and unpredictable reality that we observe. Rapid evolution is a real-time experiment that reveals the emergent and dynamic nature of resilience.

Integrating Multispecies Dynamics

Cities must evolve as ecosystems that support both human and nonhuman life. A multispecies perspective, which integrates the needs and roles of diverse species into urban planning and development, is essential for enhancing urban resilience. This approach recognizes that cities are dynamic, shared environments where various species interact, cohabit, and coevolve, contributing to the stability and adaptability of the urban ecosystem.

By understanding and incorporating the ecological roles, evolutionary paths, and survival strategies of a wide array of species—plants, animals, microbes, and humans—cities can more effectively adapt to and thrive amid environmental uncertainties. This requires seeing the world through the eyes of other species. It involves understanding how different species respond to urban pressures, how they interact with each other, and how their interactions shape the ecosystem functions and well-being. Embracing this perspective will enable us to harness organisms' evolutionary strategies to build resilience.

For instance, urban ecosystems that actively support a diversity of species can improve critical ecosystem functions such as pollination, climate regulation, and pest control. Fostering symbiotic relationships between humans and other species (e.g., designing green spaces that benefit both biodiversity and human health) can further enhance these functions, creating a more resilient and interconnected urban environment. Moreover, considering the evolutionary potential of species in urban environments can guide strategies to maintain genetic diversity, ensure species persistence, and foster the development of adaptive traits that reinforce ecosystem stability.

A perspective that values the interconnectedness of all life forms in the city is crucial for creating urban environments that are not only flexible and adaptive but also capable of sustaining the dynamic relationships that drive eco-evolutionary processes. A multispecies approach is, therefore, a fundamental strategy for building cities that are resilient, adaptable, and equipped to thrive in an increasingly unpredictable world.

Harnessing Cities' Eco-Evolutionary Potential

Cities across the world are investing significant resources in planning for resilience. Embracing an eco-evolutionary perspective will require strategies to protect and restore natural areas, such as parks, forests, and wetlands, not only to provide habitat for a wide range of species but also to preserve genetic diversity. Rethinking climate change adaptation strategies may involve innovative engineering solutions, such as designing floating communities that coexist with inevitable floods by integrating natural processes and species traits into their construction. By expanding the role of blue and green infrastructure—including green roofs, rain gardens, and permeable pavements—cities can create habitats for wildlife while also supporting species adaptation to urban environments. Land-use practices must strike a balance between urban density goals and the need for landscape heterogeneity and connectivity, integrating green spaces and natural corridors to support the conservation of species and genetic diversity. Additionally, transportation systems that reduce greenhouse gas emissions and improve air quality can also support species adaptation by mitigating pollution and climate change. Supporting local food systems, such as community gardens, farmers' markets, and urban agriculture, further contributes to the genetic diversity of crops and animals by promoting a broad range of plant and animal varieties.

Urban biodiversity conservation fundamentally alters our understanding of how to plan and design cities. By recognizing how urban environments shape species traits, designers and planners can simultaneously mitigate the impacts of urbanization and climate change while enhancing cities' adaptive capacity to include diverse communities of both people and other organisms. An evolutionary perspective helps us appreciate how historical system dynamics have shaped a system's capacity for adaptation and the evolutionary potential of organisms.

This perspective redirects planning toward fostering human–nature interactions, adaptive feedback mechanisms, and flexible institutional settings[48] that enable new forms of cooperation between humans and the biosphere.[49] Viewing cities through an eco-evolutionary lens broadens the dimensional space in which human–nature relations unfold, revealing potential pathways for a new coexistence. Rather than imposing predefined "solutions" for communities to implement, planning and design should be guided by principles of resilience, innovation, and evolvability of complex systems.[50]

Complex systems exhibiting greater *heterogeneity* and *modularity* tend to have higher adaptive capacity compared to systems dominated by connectivity and homogeneity. Other attributes such as *cross-scale interactions, early warning* mechanisms, and *self-organization*, further enhance their potential for adaptation and evolutionary innovation.[51]

Incorporating these principles into urban design and planning strengthens the evolutionary resilience of populations and landscapes, safeguarding critical ecosystem functions on which the city depends (figure 4.2).[52] This approach advances urban sustainability by achieving key evolutionary objectives: maintaining genetic diversity, facilitating evolutionary potential, aiding species with evolutionary constraints, identifying and protecting evolutionary refugia, enhancing gene flow, and bolstering adaptability to future environmental change.[53]

The inherent *heterogeneity* of urban landscapes creates diverse microhabitats, thereby boosting the evolutionary potential of multiple species and genotypes.[54] Designing cities from an eco-evolutionary perspective allows different genotypes to evolve traits suited to fluctuating environments.[55] For instance, a variety of plant species in urban gardens can drive the evolution of growth and defense traits better adapted to urban conditions.[56] This heterogeneity also supports species with limited evolutionary adaptability, such as the white-footed mouse in New York's urban parks.[57] The diverse habitats can serve as evolutionary refugia, offering sanctuaries for unique species.[58] In cities, these refuges might include vacant lots and gardens that can support species preadapted to the unique stresses of urban life, such as low-nutrient soils or drought conditions.[59]

Principles of Urban Eco-Evolutionary Resilience

	Heterogeneity	Modularity	Cross-Scale Interactions	Early Warning	Self-Organization
Genetic Diversity	Heterogeneous urban landscapes support diverse species and genotypes.	Differential selection in modular populations maintains multiple genetic lineages.	Interactions across scales create opportunities for genetic exchange and diversification.	Monitoring genetic indicators enables early detection of changes in diversity and population condition.	Local interactions generate spatial genetic structure, forming distinct genetic groups.
Evolutionary Potential	Environmental variation exposes populations to differing selective pressures that expand adaptive potential.	Semi-independent population units allow adaptive variation to accumulate under distinct conditions.	Cross-scale feedbacks expose populations to multi-level selective pressures, broadening adaptive pathways.	Early signals identify emerging traits with adaptive potential in urban environments.	Self-organizing dynamics produce novel combinations of traits and genotypes.
Genetically Constrained Mismatch	Heterogeneous conditions buffer low-evolutionary-potential species by enabling localized adaptation.	Modular structure limits perturbation diffusion, stabilizing populations during environmental shifts.	Cross-scale demographic and genetic exchanges stabilize populations with limited evolutionary capacity.	Early detection of lineage-specific stress improves understanding of mismatch risk in constrained species.	Self-organizing dynamics support persistence in species with limited evolutionary capacity.
Evolutionary Refugia	Variation in habitat structure creates pockets of reduced disturbance that serve as evolutionary refugia.	Modular habitat structures create semi-isolated conditions that preserve unique evolutionary trajectories.	Cross-scale exchanges allow refugial genetic variation to contribute to broader adaptive dynamics.	Signals of localized evolutionary distinctiveness help identify emerging refugial areas.	Self-organizing habitat patches create conditions in which rare lineages can persist.
Gene Flow	Spatial heterogeneity facilitates gene flow across diverse habitats, preserving genetic diversity.	Modular landscapes create selected connectivity that shapes gene flow among populations.	Cross-scale interactions enable genetic exchange within and among meta-populations.	Disruptions to gene exchange can be detected through genomic and movement indicators.	Self-organizing dynamics shape movement and interaction patterns that support gene flow.
Future Socio-Ecological Adaptability	Diverse environmental conditions maintain a broad repertoire of alleles relevant to future extremes.	Modularity maintains distinct population responses, sustaining adaptive potential under future conditions.	Cross-scale exchanges maintain genetic variation that enhances future adaptive potential.	Genetic signals serve as indicators of populations' capacity to respond to future environmental pressures.	Self-organizing responses diversify adaptive trajectories, sustaining long-term evolutionary flexibility.

Figure 4.2
Principles of urban eco-evolutionary resilience. Source: Author-created image.

Urban design can further enhance eco-evolutionary feedback by adopting a *modular approach*, which recognizes the importance of differential selection pressures and the emergence of local adaptations.[60] Modular landscape design enhances population stability and robustness by buffering against perturbations and creating semi-independent compartments that sustain different species and ecological interactions.[61] This approach also facilitates the conservation of "evolutionary refuges" and promotes gene flow between modules, thereby preserving genetic diversity within the population.[62]

A *cross-scale design* approach is crucial in urban planning. By considering functional redundancy across scales and encouraging genetic exchange between populations, urban environments can maintain and even enhance genetic diversity.[63]

Monitoring changes and implementing *early warning systems* in these diverse and interconnected environments is also essential for detecting the emergence of new adaptive traits or disruptions in gene flow, thereby allowing for timely responses to changing environmental pressures.[64]

Furthermore, fostering *self-organization* within these environments enables natural and social systems to adapt internally to external changes, supporting the emergence of novel traits and facilitating the evolution of threatened taxa.[65]

Implementing principles of eco-evolutionary resilience transforms cities and their infrastructure into assets that maintain and enhance evolutionary potential, enabling species to adapt to present and future rapid environmental changes.

To reframe urban planning from an eco-evolutionary perspective, we must shift our focus from asking what is best for a planet without humans to considering what is best for a planet inhabited by people. What may still surprise some is the realization that cities are home to a vast array of plant and animal species, many of which are endangered, and that these organisms have developed strategies to coexist with us. As explored in the remainder of this book, rethinking our cities in light of eco-evolutionary change will require us to imagine and evolve a new capacity for bio-cooperation between humans and other species.

5 Emergent Properties

Billions of years ago, microscopic cyanobacteria initiated a planetary transformation. As they produced oxygen through photosynthesis, their by-products began to accumulate in the atmosphere, setting off a chain reaction that reshaped Earth. This wasn't just a simple chemical shift—it was the catalyst for evolutionary leaps that would define the planet's future. As oxygen levels rose, life adapted, and organisms capable of breathing this new air began to emerge. The real breakthrough came when early eukaryotic cells engulfed cyanobacteria, forming a symbiotic relationship that would change the course of evolution. Over time, these cyanobacteria evolved into chloroplasts, enabling their hosts to harness the power of sunlight. This cooperation between once-independent organisms fueled the rise of complex ecosystems and biodiversity.[1]

Today, cities stand on the verge of a similar evolutionary leap. Like cyanobacteria billions of years ago, life in cities must adapt to a rapidly changing climate—yet this time, survival hinges on the cooperation between humans, nature, and technology. Just as the cyanobacteria transformed life on Earth through symbiosis, humans and nonhuman species in cities have the potential to transform the future of an urbanized Earth through bio-cooperation to endure the challenges of heat, rising seas, wildfires, and water scarcity.

In the past, urban development often prioritized human needs, progressing from ancient settlements that coexisted with nature to industrial cities that increasingly isolated themselves from natural systems. But just as evolutionary processes have shaped life on Earth, the future of cities will depend on the evolution of cooperative interactions among species, systems, and technologies. This new form of urban symbiosis could reshape cities into thriving ecosystems where humans and other life forms coexist, driving a more resilient, sustainable evolution of our urban environments. Just as the symbiotic relationships between early eukaryotes and cyanobacteria gave rise to new forms of life, bio-cooperative cities can foster new models of urban resilience, in which survival is no longer rooted in domination over nature, but in collaboration with it.

By envisioning cities through the lens of bio-cooperation, we unlock the potential to reshape contemporary urban landscapes. Looking to the future offers a new perspective—one in which multispecies cooperation initiates a new phase in urban evolution, increasing complexity and adaptive capacity by reestablishing connections between humans and nature. Much as cyanobacteria once transformed Earth's atmosphere and made new life systems possible, bio-cooperative cities may help urban environments evolve into more adaptive and enduring forms, where resilience emerges from reciprocity with ecological processes rather than separation from them.

Emergence of Bio-Cooperative Cities

Fast forward to 2080. The alarming climate change projections once reserved for 2100 have already become a reality. Coastal cities across Southeast Asia—such as Jakarta, Bangkok, and Ho Chi Minh City—find themselves on the front lines of an intensifying climate crisis. Sea levels have risen dramatically, reaching the highest end of the Intergovernmental Panel on Climate Change (IPCC) scenarios at 1 meter, leaving these urban centers vulnerable to relentless flooding and storm surges. What was once considered a distant threat has become a daily battle for survival, exacerbated by the region's rapidly urbanizing population. These cities face not only the physical encroachment of the ocean but also widespread ecological degradation that threatens their very existence.

In desert cities like Phoenix and Dubai, and in regions with unpredictable monsoon climates like Mumbai, temperatures have reached unprecedented extremes, with summer highs consistently surpassing 50°C (122°F). Cities like Athens, Sydney, and Los Angeles have become synonymous with the devastating impacts of wildfires. The increasing frequency and intensity of these fires—fueled by rising temperatures, prolonged droughts, and high winds—have turned these cities into battlegrounds where human civilization and nature collide with terrifying consequences. In Athens, with its rich history and proximity to dense forests, fires have engulfed entire neighborhoods, destroying both ancient ruins and modern infrastructure. Sydney, surrounded by vast bushlands, has faced wildfires that stretched for thousands of kilometers, decimating wildlife and threatening suburban communities. Los Angeles, already familiar with wildfires, now faces an unprecedented crisis, as climate change, urban sprawl, and persistent drought turn the hills and valleys of Southern California into a perpetual fire hazard.

By 2080, cities like Cape Town, São Paulo, and Lima will have become global symbols of the fight against water scarcity. Erratic rainfall patterns, prolonged droughts, and the mismanagement of water resources have pushed these cities to the brink, forcing them to confront the harsh realities of life in a water-scarce world. Cape Town, which narrowly averted "Day Zero" in 2018, the point at which municipal water supplies would have been shut off, faces even more severe challenges as climate change exacerbates the frequency and intensity of droughts. São Paulo, dependent on a complex network of reservoirs and rivers, struggles to maintain its water supply as deforestation in the Amazon Basin disrupts rainfall patterns. Lima, located in one of the world's driest coastal deserts, has seen its glacial meltwater sources dwindle, threatening the city's very survival.

In the years leading up to this point, various strategies were implemented to protect coastal cities from rising seas. Populations were relocated inland, experimental floating cities were constructed to rise with the tides, and shorelines were fortified with elevated roads and natural barriers. Large-scale projects to restore wetlands and replant mangroves were widely adopted, grounded in the understanding that nature could play a crucial role in buffering against climate impacts. Salt marshes filtered pollutants, stabilized sediments, and provided unique habitats for wildlife, while seagrasses contributed to water clarity and carbon sequestration and served as crucial nursery grounds for marine species. Mangrove forests, valued as both natural barriers against storm surges and critical carbon sinks, were central to these efforts. However, many of these initiatives were carried out in isolation, lacking a comprehensive understanding of the intricate ecosystems they aimed to protect.

Early adaptation efforts in cities facing unbearable heat focused on technological solutions. Phoenix invested heavily in water conservation, the construction of massive desalination plants, and the development of "cool roofs" and green walls designed to reduce urban heat. Mumbai launched ambitious water management projects, including the rejuvenation of its lakes and rivers, alongside the construction of desalination facilities to supplement its freshwater supply. Dubai, already a leader in cutting-edge technology, pursued similar strategies, emphasizing desalination and the construction of artificial oases and cooling infrastructure. However, as temperatures continued to rise and water sources became increasingly scarce, the limitations of these technocentric approaches became apparent.

In response to the devastating wildfires, Athens, Sydney, and Los Angeles focused on the most innovative firefighting technologies and infrastructure. Firebreaks were constructed, aerial firefighting units were deployed, and early warning systems were established. However, these measures, while effective in the short term, could not keep pace with the escalating fire risks driven by climate change. The fires grew more frequent and destructive, and the human and ecological toll mounted. It became clear that a new approach was needed—one that not only mitigated fire risk but also restored the region's ecological resilience.

A similar technological optimism characterized the immediate responses of the cities facing water scarcity and drought. Cape Town, São Paulo, and Lima initially turned to large-scale engineering projects, such as the construction of new dams, desalination plants, and river diversions. However, these interventions soon proved inadequate as the scale of the water crisis continued to grow. The cities were forced to rethink their approach, recognizing that true resilience would require not just technological solutions but also a commitment to restoring and protecting their natural water sources.

The Awakening

As cities reflected on early twenty-first-century climate adaptation strategies, driven by urgency but often constrained by fragmented, short-sighted approaches, the limitations of these approaches became evident. Initial optimism waned as the shortcomings of one-size-fits-all solutions were exposed. These efforts, while well-intentioned, revealed critical limitations that underscored the need for a deeper understanding of ecological and social complexity.

In coastal cities, early mangrove restoration efforts often relied on monoculture plantations of fast-growing species. These simplified ecosystems, while offering immediate benefits, like short-term erosion control, lacked the resilience of diverse, natural ecosystems and became vulnerable to pests, diseases, and environmental changes, leading to high mortality rates. The lesson became clear: Ecosystem restoration requires a deep understanding of the complex dynamics that sustain these environments over time.

Efforts to restore wetlands further highlighted the importance of working with the underlying hydrological processes. Restoration projects often focused on replanting vegetation and reconstructing landforms—visible and immediate actions—while sometimes overlooking the critical role of habitat quality and natural water flow patterns. This led to challenges such as water stagnation, poor oxygenation, pollution, and disrupted sediment deposition, which ultimately resulted in wetlands that lacked long-term resilience.

Floating cities, heralded as cutting-edge solutions to sea-level rise, revealed significant flaws. Challenges such as disruptions to marine ecosystems, altered water quality, and accessibility for vulnerable populations called into question the sustainability of these innovations. These challenges highlighted the urgent need for extensive research and monitoring to ensure the viability of such innovations, while also emphasizing the importance of inclusive and equitable solutions.

Urban heat offered another stark reminder of the pitfalls of overly simplistic solutions. While tree-planting campaigns were widespread, many failed to consider local conditions or species, limiting their effectiveness. Fire-prone regions, meanwhile, suffered as their focus remained on suppression rather than prevention. Sydney's devastating loss of biodiversity underscored the need for rewilding programs to restore fire-resistant native species and enhance natural firebreaks.

Beyond these specific interventions, broader environmental and societal challenges further complicated restoration efforts. Persistent stressors like pollution, overfishing, and unsustainable land use continued to degrade ecosystems, weakening the resilience of both natural and engineered solutions.

These missteps highlighted a broader truth: Isolated interventions, divorced from the complexities of ecosystem dynamics and social equity, were inadequate. Addressing urban resilience required a holistic, adaptive approach—one that considered ecological function, biodiversity, and human well-being as interconnected priorities. The awakening came from recognizing these shortcomings—and the necessity of moving beyond them.

A Strategic Shift

By 2080, it became evident that fragmented, short-sighted interventions of the past were misguided. The escalating climate crisis spurred a collective rethinking of humanity's relationship with nature. A new strategy emerged—grounded in ecological complexity and informed by the interconnectedness of all species. Urban development began to be driven by principles of coexistence and mutual benefit, acknowledging that humans are integral to, rather than separate from, ecological systems.

Multispecies cooperation emerged as a crucial pathway to resilience. This approach reimagined cities as coevolving ecosystems in which the interplay among species became integral to their adaptability. Urban planning shifted from a narrow focus on human-centric needs to a broader framework that incorporated the ecological functions and needs of diverse species. By aligning human development with multispecies dynamics, planners created urban systems that responded effectively to environmental challenges.

Building on this ecological foundation, urban design began to account for eco-evolutionary dynamics. Species were chosen for their functional traits and adaptive potential under shifting conditions. Maintaining genetic variation, enabling dispersal, and supporting trait change over time became integral to sustaining ecosystem performance.

Equity became a central design criterion. Planners recognized that resilience could not be achieved without addressing structural social inequalities in access to green space, clean water, and climate protection. Interventions were prioritized in underserved areas, and urban systems were redesigned to redistribute ecological benefits and reduce vulnerability across social groups.

A new "urban commons" emerged, where citizens, scientists, and policymakers collaborated to design cities that support species essential to ecosystem resilience. This vision redefined nature-based solutions by leveraging the adaptive capacities and functional diversity of species. For instance, coastal cities implemented diverse mangrove ecosystems designed to mirror the complexity of natural habitats and incorporate species with complementary functions. By restoring hydrological cycles and considering species-specific traits, cities developed ecosystems better equipped to withstand disturbances. By considering different response and effect traits—an approach referred to as the "portfolio effect" or insurance hypothesis—these ecosystems are more likely to maintain their functions in the face of disturbances, thereby strengthening ecosystems and enhancing urban resilience.

Wetland restoration shifted toward reactivating natural hydrological processes, with attention to flow regimes, salinity gradients, and sediment transport. As climate pressures intensified, cities applied the new strategy across ecosystems—integrating ecological processes, trait-based species selection, and spatial equity. In arid regions, this meant redesigning urban systems to manage heat, water scarcity, and social vulnerability together.

In Phoenix, nature-based solutions were deployed to increase shade and create cooler microclimates, drawing on species and spatial forms inspired by desert ecosystems. Disadvantaged neighborhoods were prioritized, aligning climate adaptation with equity goals. The city reintroduced natural floodplains and developed water-harvesting systems to reduce dependence on external supplies. Drought-tolerant species such as *Prosopis velutina* (velvet mesquite), *Parkinsonia florida* (blue palo verde), and *Encelia farinosa* (brittlebush) were selected for their trait variation and adaptive potential, ensuring long-term performance under heat and water stress.

Mumbai shifted focus to reviving traditional water management practices, such as rainwater harvesting and wetland restoration. Reforestation with native trees provided shade, absorbed excess carbon, and cooled the urban landscape. These efforts were coupled with initiatives to improve water access in underserved neighborhoods, ensuring that the benefits of reforestation and urban cooling were equitably distributed. Dubai, acknowledging the fragility of its water-intensive development model, integrated its urban infrastructure with the surrounding desert ecology. This included protecting natural desert landscapes, promoting xeriscaping, and developing urban green spaces mimicking the Arabian desert's flora and fauna.

Cities prone to wildfires shifted their approach from combating fires to restoring nature's capacity to prevent them. In Athens, resilience efforts focused on integrating traditional fire management practices with modern ecological restoration techniques. Controlled burns, a practice long used by Indigenous Peoples, were reintroduced to reduce flammable underbrush. Native Mediterranean species such as *Quercus coccifera* (kermes oak), *Phlomis fruticosa* (Jerusalem sage), and *Pistacia lentiscus* (mastic tree)—fire-adapted and important for post-fire recovery—were restored to maintain biodiversity.

Sydney embarked on an ambitious rewilding program to restore native bushlands and reintroduce species playing vital roles in natural fire management. Species such as *Eucalyptus punctata*, *Banksia serrata*, and *Themeda triandra* were prioritized for their fire-adapted traits, regenerative capacity, and ability to maintain biodiversity under changing fire regimes. The city also promoted the creation of green corridors, networks of parks, and natural reserves serving as firebreaks and wildlife habitats, enhancing overall resilience to fires and other climate impacts.

Los Angeles adopted a multifaceted approach to wildfire resilience, integrating urban planning with fire management. The city promoted fire-resistant landscaping and encouraged the construction of fireproof buildings in high-risk areas. By collaborating with Indigenous communities, traditional fire stewardship practices such as controlled burns were reintroduced to reduce fuel loads and maintain ecological balance. These cultural burns have been recognized for their effectiveness in mitigating wildfire risks and promoting forest health. Simultaneously, Los Angeles invested in large-scale ecological restoration projects, including restoring surrounding hills with native species such as *Quercus agrifolia* (coast live oak) and the fire-adapted shrub *Adenostoma fasciculatum* (chamise), restoring natural watercourses, and preserving open spaces that act as natural fire buffers, thereby enhancing the city's wildfire resilience.

In response to water scarcity, cities reevaluated water management strategies, focusing on restoring and protecting natural water sources. Cape Town prioritized rehabilitating watersheds and wetlands, recognizing their critical role in water filtration, groundwater maintenance, and drought mitigation. Ambitious projects were launched to restore river systems, protect natural areas, and reforest slopes with native vegetation, securing water supplies and providing habitats for local wildlife, thereby contributing to ecosystem resilience. Genetically diverse native populations—such as *Leucadendron salignum* and *Rapanea melanophloeos*—were favored to maintain evolutionary responses to fluctuating rainfall and temperature regimes.

São Paulo, recognizing deforestation's impact on water supply, implemented strict regulations to prevent further deforestation and launched reforestation projects to restore degraded areas. Parallel investments in green infrastructure, such as urban wetlands and rain gardens, enhanced the city's ability to capture and store rainwater, reducing reliance on external sources and increasing resilience to future droughts. Species introduced into these systems—such as *Schinus terebinthifolia* (Brazilian peppertree), *Cecropia pachystachya*, and *Inga marginata*—were selected for functional complementarity and adaptive potential ensuring ecological performance under a range of future conditions.

Lima, facing the loss of glacial meltwater, turned to ancient water management techniques developed by pre-Columbian civilizations. The reintroduction of traditional irrigation systems, such as *amunas* (pre-Incan water channels) and terraced agriculture, combined with modern water-saving technologies and wetland restoration, allowed Lima to better manage scarce water resources and reduce vulnerability to drought.

Technology plays an integral role in this new strategy, enabling real-time monitoring of ecosystems through biosensors and artificial intelligence (AI). AI models predict how specific plant species might respond to rising salinity, heat, flooding, and other environmental stressors, guiding city planners in selecting optimal species mixes for coastal protection. Advanced biosensors, harnessing the sensory capacities of various species, monitor environmental conditions in real time and prevent ecological collapse. These sensors, often deployed via drones, enable rapid adjustments to conservation strategies, ensuring interventions are timely and effective.

This emerging strategy presents both opportunities and challenges. Transitioning from traditional, human-centered urban planning to a more inclusive, multispecies approach requires significant shifts in thinking about what a city is and about the practices of urban design and planning. It also raises new scientific questions about how species perceive the environment, which can expand our ability to experience the world through their sensory systems for detecting light, sound, humidity, airborne chemicals, and other stimuli beyond human perception. Cities must adopt flexible, adaptive mindsets, embracing uncertainty and experimenting with new ideas. This also demands ongoing collaboration and knowledge-sharing, both within and between cities. Recognizing the need for coordinated efforts, cities across Southeast Asia have begun forging stronger connections, creating networks for sharing data, strategies, and innovations. These networks enable cities to learn from each other, avoid past mistakes, and build on successful strategies, fostering a shared purpose and collective action essential for addressing the complex challenges posed by climate change.

Emergent Properties of Bio-Cooperative Cities

Bio-cooperative cities arise from the evolution of interspecies cooperation as an emergent property of eco-evolutionary feedbacks. Their distinctive qualities are shaped by dynamic interactions between humans and other species, where cooperation serves as both an adaptive strategy in the face of environmental change and a mechanism for resilience—shaped by the contingent pathways of evolutionary processes.

This paradigm builds upon but moves beyond earlier models such as biophilic and regenerative cities. Biophilic cities emphasize nature's restorative role in reconnecting humans with the natural world,[2] while regenerative cities envision urban spaces as instruments of ecological renewal.[3] Bio-cooperative cities, by contrast, center on symbiotic relationships among species, highlighting the coevolution of human and nonhuman life within the urban fabric. These mutualistic interactions foster resilience and adaptability by enabling novel forms of coexistence to emerge.

The focus shifts from simply incorporating natural elements to intentionally designing cities that support active cohabitation and cooperation among humans and other species. Bio-cooperative cities recognize urban spaces as shared habitats where species interact in mutually beneficial ways, adapting together to the challenges of an urbanizing planet. These cities are defined by a deliberate acknowledgment of the agency of all urban inhabitants—human and nonhuman and an explicit recognition of the ecological requirements of diverse life forms, reflected in the evolving niches they occupy—the roles they perform, the resources they rely on, and the relationships that sustain them. This perspective transcends enhancing human well-being or conserving biodiversity by striving to create equitable urban spaces where all species contribute to—and depend on—the resilience and adaptability of the urban ecosystem.

Bio-cooperative cities are grounded in eco-evolutionary dynamics, biological complexity, adaptive feedbacks, and institutional flexibility, which support continuous cooperation between species and systems. By fostering symbiotic relationships between humans and the biosphere, bio-cooperation reframes urban resilience as an emergent, coevolutionary property of urban systems. It redefines planning and design as responsive, situated practices that embrace the complexity and heterogeneity of urban life.

Rather than providing predetermined solutions, this approach advances principles that enhance adaptive evolution and strengthen cities' transformative capacity. Bio-cooperative cities are designed to respond dynamically to interconnected challenges—climate change, biodiversity loss, social inequity, and the sustainable management of essential resources such as water, food, and energy. Through bio-cooperation, cities can evolve into ecosystems that not only sustain life but also amplify the resilience and adaptive capacity of all species.

Evolving Urban Niches

Urban ecosystems are composed of diverse species that occupy shared and specific ecological niches, each contributing to the overall function and stability of the environment.[4] In a bio-cooperative city, the design process begins with a rigorous analysis of these niches, identifying how various species—such as pollinators, decomposers, and apex predators—interact within the urban matrix. This principle recognizes that the survival and flourishing of these species are contingent on the availability of suitable habitats and resources, which can be enhanced through strategic urban planning. For example, green roofs and vertical gardens can serve as microhabitats for pollinators, while constructed wetlands can support amphibian populations and water filtration functions. The goal is to create a spatial configuration that integrates natural and built environments, allowing for the continuous interaction and coevolution of species within the urban landscape.

In contrast to traditional urban restoration efforts that often isolate nature within parks or preserves, the bio-cooperative approach views the entire city as a habitat mosaic in which ecological processes are intertwined with urban infrastructure. By fostering interspecies symbioses—for example, mutualisms between urban trees and mycorrhizal fungi—cities can enhance ecosystem functioning including carbon sequestration, soil stabilization, and temperature regulation. These symbiotic interactions not only support biodiversity but also contribute to the resilience of urban ecosystems in the face of environmental changes. For instance, the adaptive foraging behaviors of birds in urban environments can help control insect populations, reducing the need for chemical pesticides. By designing and maintaining urban spaces that offer birds food sources, shelter, and nesting sites, we can actively support bird populations and promote reciprocal relationships within the ecosystem. This principle underscores the importance of creating cities that enable the coexistence of diverse species, ensuring that urban ecosystems remain functional and resilient.

Eco-Evolutionary Dynamics

Eco-evolutionary dynamics are a defining property of bio-cooperative cities. Urban environments impose distinctive selective pressures—heat, fragmentation, altered hydrology, pollution, and novel resource regimes—that drive evolutionary change in resident species. As traits shift, species interactions and ecosystem processes can shift with them, influencing biodiversity, seed dispersal, primary productivity, and water and air purification. In this way, eco-evolutionary change becomes part of the operating logic of the city, shaping ecological function on contemporary time scales.[5]

In bio-cooperative cities, this property shapes how cities are designed, governed, and maintained over time. Bio-cooperative cities recognize that the long-term functionality of nature-based solutions,[6] increasingly implemented as cost-efficient strategies to enhance urban resilience, depends on populations that can evolve. Adaptive evolution in key traits can undermine nature-based solutions—for example, pest resistance to biological control, shifts in flood-control vegetation that reduce root stability, and changes in plant traits that affect carbon uptake and pollutant mitigation.[7] Eco-evolutionary feedback can also strengthen nature-based solutions by expanding tolerance to urban stressors and stabilizing key ecological controls. In urban lakes and ponds, *Daphnia* has evolved tolerance to cyanobacteria, warming, and road salt, enabling it to suppress phytoplankton and reduce the risk of harmful blooms—supporting water-quality regulation under chronic urban stress.[8]

Designing with eco-evolutionary dynamics in mind means viewing nature-based solutions as living systems whose resilience depends on maintaining evolutionary potential through diversity, connectivity, and selection environments that support functional stability.

Multispecies Equity

Multispecies equity challenges the anthropocentric bias in urban planning by acknowledging the intrinsic value and agency of nonhuman species.[9] In a bio-cooperative city, urban spaces are designed not only for human benefit but also to meet the needs and protect the rights of other species that share the environment. Bio-cooperative cities protect migratory bird nesting habitat through seasonal disturbance buffers and retrofit-safe building guidelines, secure reliable clean-water access for urban wildlife through distributed ponds, bioswales, and contamination controls, and maintain habitat connectivity through green corridors that support movement, foraging, and life-cycle completion. This design approach is grounded in understanding the ecological roles and life histories of urban species, as well as the social and cultural dimensions of human–nature relationships. Designing for multispecies equity draws on ecological justice frameworks, which argue that all species have a right to the resources necessary for their survival and well-being.

Prioritizing multispecies equity in urban design involves creating governance structures that incorporate the perspectives of nonhuman species, often informed by Indigenous ecological knowledge and participatory approaches.[10] For instance, urban planning could include citizen-led biodiversity monitoring programs that track the health of local species populations and inform management decisions. Additionally, policies could be enacted to protect critical habitats and migratory corridors from development, ensuring that urbanization does not disproportionately harm vulnerable species. By prioritizing multispecies equity, cities can foster more inclusive and resilient ecosystems where the well-being of all inhabitants—human and nonhuman alike—is actively supported and protected. This principle not only enhances biodiversity but also contributes to the cultural and ecological richness of urban environments.

Evolvability, Adaptability, and Foresight

Bio-cooperative cities exhibit the interdependent properties of evolvability, adaptability, and foresight, which enable them to remain functional in the face of accelerating environmental change and increasing uncertainty. Together, these properties enhance long-term resilience by enabling urban systems to buffer immediate risks while maintaining evolutionary potential to adapt to future challenges.

Evolvability is the capacity of a system to generate adaptive variation, enabling it to respond to novel challenges and opportunities.[11] Evolvability in natural systems arises from the ability of populations to produce heritable variation that selection can act upon, generating novel traits over time. Applied to cities, enhancing evolvability means fostering systems that can innovate and diversify responses to change. For example, urban designs that conserve ecological functions and maintain evolutionary potential—such as green corridors supporting species dispersal—enhance the capacity of cities to adapt over generations.

In nature, evolvability complements adaptability, which provides organisms the short-term ability to cope with immediate challenges.[12] In cities, adaptability ensures that urban infrastructure remains functional amid immediate disruptions, such as extreme weather events. Features like permeable pavements or modular green roofs not only mitigate flooding but can also be reconfigured as environmental conditions shift. Foresight provides a mechanism for directing these adaptive capacities toward long-term resilience. By adopting an anticipatory approach, urban planners can reduce the vulnerability of cities to climate change and other environmental stressors while maintaining evolutionary potential to ensure that the city has the capacity to innovate and diversify its responses over time.

Designing for adaptability and foresight requires cities to embrace iterative and experimental planning processes, where continuous learning and adaptation are central. This might involve pilot projects that test new urban greening strategies or adaptive management practices that respond to real-time ecological data. By harnessing species' sensory capacities, such as physiological or behavioral responses to temperature or moisture, and their anticipatory responses, including shifts in bird migration patterns or changes in plant flowering times, cities can monitor ecological health and detect emerging changes before they escalate into crises. This proactive stance not only enhances the resilience of urban ecosystems but also ensures that cities remain vibrant and livable for future generations. By embedding adaptability and foresight into the core of urban design, cities can evolve alongside the natural world, maintaining ecological functions and services in the face of uncertainty.

Cooperative Innovation

Cooperative interactions among species are a driving force behind many of the evolutionary innovations that sustain life on Earth. Lynn Margulis's theory of endosymbiosis profoundly transformed our understanding of the evolution of eukaryotic cells (cells that have a nucleus). Protozoa, fungi, plants, and animals all resulted from partnerships involving several different prokaryotic cells (microorganisms including bacteria and archaea). Through her study of cell structure, Margulis showed that symbiosis is a fundamental driver of evolution and a source of biological novelty.

Symbiosis is an evolutionary strategy that allows species to adapt to diverse habitats. The emergence of multispecies symbiosis in bio-cooperative cities highlights its potential to enhance both urban resilience and sustainability. Bio-cooperative city planning promotes multifunctional landscapes that sustain cooperative and symbiotic relationships among nonhuman species—such as pollinators and flowering plants or mycorrhizal fungi and trees—while also enabling mutually beneficial relationships between humans and nonhuman species. By supporting multispecies cooperation, cities enhance ecosystem functioning—pollination, soil fertility, and carbon sequestration—which underpin urban biodiversity and human wellbeing. Community gardens that incorporate diverse plant species, for example, sustain pollinator–plant mutualisms while providing food, recreational space, and opportunities for social connection. These coupled human–ecological relationships strengthen urban health and resilience.

In a bio-cooperative city, human-engineered solutions play a critical role not in replacing these ecological processes but by coupling with and amplifying them to support urban resilience. Cities may deploy systems such as microbial fuel cells, which capture energy produced by bacteria decomposing organic waste, transforming waste streams into usable power within closed-loop infrastructures. Other designs integrate photosynthetic systems—such as algae-based building façades—that simultaneously regulate microclimates, capture carbon, and produce biomass. These technologies operate in concert with living systems rather than as substitutes for them, extending ecological functions into the built environment. Through this cooperative innovation, cities shift from isolated, human-centric systems into interconnected urban ecosystems that support the coevolution and persistence of human and nonhuman life.

The vision of bio-cooperative cities marks a fundamental shift in how we think about and design urban environments. Examining their emergent properties reveals core principles that guide strategies for creating dynamic, multispecies cities. By integrating shared ecological niches, harnessing eco-evolutionary dynamics, prioritizing multispecies equity, promoting cooperative innovation, and enhancing evolvability, adaptability, and foresight, we can build cities that are not only resilient and sustainable but also inclusive for all species. These principles challenge us to move beyond the narrow confines of anthropocentric urban planning to reimagine the city as a dynamic ecosystem where humans and nonhumans coexist and thrive together.

Bio-cooperative design principles can be situated within a broader continuum of strategies that integrate nature into urban systems. At one end of the spectrum are *bio-mimetic* approaches, which emulate specific forms or processes observed in nature but do not necessarily support the living systems themselves. *Nature-based solutions* move closer to ecological integration by leveraging ecosystems to provide services such as flood control, air purification, or cooling. Yet many nature-based solution implementations still rely on human-directed design and management, often overlooking the evolutionary agency of nonhuman species. *Restorative* or *conservation-based* approaches, such as urban rewilding or wetland restoration, offer greater potential for self-organized ecological processes but may remain separate from the core urban fabric.

Bio-cooperative cities differ in that they recognize urban spaces as *agents of coevolution*, where mutual adaptations between human and nonhuman organisms are ongoing and embedded in the design, governance, and function of the city. While bio-mimicry and nature-based solutions may selectively enhance survival for targeted species or functions, only strategies that support living, evolving systems—in which species retain their capacity to adapt and shape their environments—can participate fully in the evolutionary process. This distinction underscores the importance of designing urban systems that do not use nature but engage with it as a coevolving partner.

The five emergent properties of bio-cooperative cities outlined in this chapter provide the foundation for the chapters that follow, which explore the underlying dynamics that shape them, demonstrating that bio-cooperative cities are far more than an idealistic vision. They represent a viable and necessary strategy for confronting today's environmental challenges. Each chapter investigates how these properties manifest in urban ecosystems and, in so doing, begins to uncover the conditions and mechanisms through which cooperation, adaptability, and multispecies resilience are made possible. These insights form the conceptual foundation for the design principles presented in the final chapter. Though the properties are descriptive and contingent, the principles are generative—they translate observed patterns into actionable strategies for shaping cities that align with and support coevolutionary dynamics.

6 Evolving Urban Niches

In the Amazon Basin, a remarkable example of early human-driven niche construction is *terra preta*, or "black earths." Approximately 2,000 years ago, Indigenous Peoples in the region began deliberately engineering these highly fertile soils by burning organic material, such as wood and bone, and incorporating the resulting biochar into the earth.[1] This process not only enhanced the land's capacity to support agriculture, allowing for the sustenance of growing human settlements, but also altered the broader ecological dynamics of the region. Soil microbes, plant species, and other organisms evolved and adapted to the new, human-constructed niche. Terra preta remains a compelling example of human-induced ecological modification, acting as a powerful carbon sink that sequesters significant amounts of carbon for millennia.[2] It provides clear evidence that human influence, when carefully managed, can result in beneficial and long-lasting ecological impacts, challenging the simplistic view of human intervention as inherently destructive.

Beyond soil enrichment, many Indigenous agricultural practices contributed to sustaining local biodiversity. Forest gardening, for instance, involved the selective planting and management of trees, shrubs, and vines, creating a mosaic of habitats that supported both human and non-human communities.[3] Species such as *Theobroma cacao* (cacao), *Bactris gasipaes* (peach palm), and *Inga edulis* (ice cream bean tree) were cultivated, providing food not only for humans but also for fruit-eating animals such as birds, monkeys, and bats, which played a crucial role in seed dispersal.[4] Furthermore, nitrogen-fixing species like *Inga* were planted to enhance soil fertility, benefiting both agricultural crops and surrounding vegetation.[5] These strategies, combined with the creation of small forest clearings for agroforestry, encouraged a diverse understory and fostered habitats for pollinators such as bees and butterflies.

Evidence of humans as ecosystem engineers is not limited to preindustrial societies. Recent studies in the Qa' Shubayqa wetlands of eastern Jordan, dated to the Late Pleistocene, reveal that humans and migratory birds co-constructed ecosystems.[6] Through archaeological remains and biomolecular analysis of eggshells, it is evident that these wetlands, situated in a semi-arid landscape, were transformed into rich habitats for both humans and waterfowl. During the transition from the Late Pleistocene to the Early Holocene, humans likely modified the wetlands to enhance their productivity, fostering conditions that allowed migratory birds, such as ducks and geese, to breed year-round. This interaction between humans and the environment highlights the mutual construction of ecological niches. The presence of waterfowl provided a reliable food source for human populations, while the human modifications to the wetlands supported the birds' year-round habitation, a departure from their typical seasonal migration patterns.[7]

Human Niche Construction

For millennia, humans have been active agents of niche construction, reshaping both local and global biodiversity through agricultural practices, settlement patterns, and trade networks.[8] The advent of agriculture over 10,000 years ago marked a pivotal moment in human history, enabling the development of larger, more stable populations and the rise of urban settlements. Early cities, such as those in Mesopotamia, the Indus Valley, and Ancient Egypt, are among the earliest examples of large-scale urban settlements. These ancient cities began to significantly alter local ecosystems through practices like irrigation, land clearing, and the domestication of plants and animals.[9] As agricultural practices spread and populations grew, so did the ecological impacts of these ancient cities.[10] Deforestation, soil modification, and changes in water management systems affected species distributions and ecosystems, creating both disruptions and new ecological conditions.[11]

These early patterns of niche construction laid the groundwork for urbanization. Neolithic settlements like Çatalhöyük and Jericho represent proto-urban forms where agriculture, sedentism, and social complexity began reshaping local ecologies. These Neolithic settlements, while lacking the full institutional complexity of later cities, already exhibited dense habitation, material specialization, and significant ecological modification.[12] These proto-urban centers mark an early phase of human niche construction that set the stage for the emergence of Bronze Age cities.[13] But it was with the rise of industrial cities that the scale and intensity of human–environment interactions shifted dramatically.

From the Bronze Age through the early modern period, urban centers became increasingly complex and influential in shaping their environments. Preindustrial cities—such as those in Mesopotamia, the Roman Empire, and early Islamic and European polities—expanded human influence on ecosystems, though not yet at the scale of later industrial urbanism. Irrigation systems altered river courses, forests were cleared for construction and fuel, and urban expansion fragmented surrounding landscapes. These transformations disrupted local ecological processes, shifting species distributions and changing water and nutrient flows.

Yet many preindustrial cities also engaged in deliberate niche construction that fostered multispecies coexistence. For instance, cities like Amsterdam and Venice developed intricate canal systems that managed water levels for human use but also created new aquatic habitats. In Japan, rice paddies designed for agriculture also supported fish and migratory birds, inadvertently enhancing local biodiversity.

The transition from early agrarian settlements to industrial and postindustrial cities marks a profound shift in humanity's capacity to modify ecological systems. The onset of industrialization in the late eighteenth and nineteenth centuries transformed cities into centers of mechanized production and commerce. Resource demands extended far beyond local ecosystems, fueling deforestation, pollution, and habitat fragmentation on a global scale. Cities like Manchester, London, and New York became industrial hubs powered by coal, railroads, and factories—driving widespread environmental degradation. This period reinforced a stark separation between urban environments and the natural systems that once sustained them, embedding a dichotomy between cities and nature.

Yet the human modification of ecosystems did not begin with industrialization. Early agricultural societies were already ecosystem engineers—using fire to clear land and developing irrigation systems to support crop cultivation. Over time, these practices gave rise to novel ecosystems and anthropogenic biomes. Urbanization has dramatically intensified the scope and speed of ecological transformation. What has emerged with modern urban expansion is not only a new human habitat but a new mode of niche construction—one that transcends geographic boundaries. Today, cities operate as global nodes for the movement of species, materials, and ecological processes, reshaping the evolutionary trajectories of both human and nonhuman life. These processes continue to redefine urban ecosystems and their implications for biodiversity.

As urbanization expands, the niches of many species have been displaced, disrupted, or fundamentally altered. Species once common in rural or wild habitats are now forced to adapt to urban conditions—or face extinction. Some, like pigeons, rats, and certain weedy plants, have thrived by exploiting human-made environments and resources. Others, especially large predators and forest-dependent species, have been excluded from urban areas, their ecological roles diminished or lost altogether.

By the nineteenth and twentieth centuries, urbanization had transformed cities into human-dominated landscapes. Infrastructure—roads, sewers, power grids—fragmented habitats, while green space was often isolated within parks rather than woven into the urban fabric. This spatial separation reinforced the view of cities as exclusively human domains, with nature pushed to the margins and treated as something to be managed. The shift from rural, agrarian societies to industrialized urban centers redefined the human niche, with far-reaching consequences for biodiversity and ecological functioning.

The long-term dynamics of sociocultural niche construction are now visible in the Anthropocene—an epoch in which human activity has become a dominant force shaping Earth's ecological systems. Although cities occupy only a small portion of Earth's surface, they rely on resources from distant regions to sustain urban life and the activities of their inhabitants, extending their ecological impact beyond their physical footprint. Our exceptional ability to innovate and utilize technology has vastly extended the limits of the human habitat, elevating our species to the position of planetary engineers. What has evolved with the emergence of urban settlements and rapid global urbanization is a *new human niche*. In this process, humans have transformed niche construction of other species by changing important community organizing forces, such as habitat modification, predation, and resource variability.[14]

What enabled humans to scale their environmental influence so dramatically, reshaping ecosystems across the globe?

Sociocultural niche construction theory suggests that this upscaling of human influence is rooted in our ability to socially and culturally transmit knowledge about environmental management across generations, accelerating the pace of ecological change. Sociocultural niche construction theory holds that humans shape and modify their ecological environments through social and cultural processes. Unlike other species that adapt to environmental constraints, humans actively alter their surroundings through learned behaviors, technological innovation, and social cooperation. Cultural traits have overwhelmed genetic traits in shaping the human niche since they evolve more rapidly than genetic traits due to horizontal transmission—the ability to spread within a single generation through social learning and communication—allowing for swift adaptation, while the ratchet effect further accelerates this process by enabling the accumulation and combination of cultural innovations, leading to increasingly complex institutions and technologies across generations.

Dominant Narratives

Prevailing explanations for the domination of nature and environmental degradation often emphasize humanity's unparalleled evolved capacity to shape its environment through sociocultural niche construction.[15] These explanations focus on specific historical and evolutionary developments in human societies.

One of the most compelling explanations is the theory of *historical contingency*, which posits that human domination over nature was not an inevitable outcome but rather the result of particular historical developments.[16] The advent of agriculture serves as a critical turning point in this narrative. As human societies transitioned from foraging to farming, the need to produce surplus food for growing populations intensified resource extraction and habitat modification.[17] This shift catalyzed the development of hierarchical social structures, which, in turn, drove competitive pressures and cycles of increasing environmental manipulation. From this perspective, agricultural—and later, industrial—societies became locked into a trajectory of maximizing energy and resource use, often at the expense of ecological systems.[18]

Another key explanation centers on the evolution of *competitive sociopolitical structures*. As human societies grew in complexity, they began competing for resources, land, and power. This competition incentivized aggressive environmental control and exploitation. In hierarchical societies, especially those dominated by empires or capitalist systems, leaders had a vested interest in extracting as many resources as possible from the environment to maintain power and influence. This pattern of environmental domination, driven by the need for political and economic expansion, became deeply ingrained in human development. Under these systems, nature was increasingly seen as a resource to be harnessed and controlled to sustain economic growth.[19]

Cultural and ideological shifts also played a significant role in shaping human attitudes toward the environment. The rise of dualistic thinking—especially in Western societies—positioned humans as separate from, and superior to, nature. This worldview, rooted in philosophical and scientific traditions emerging from the Scientific Revolution and the Enlightenment, encouraged the belief that nature could be controlled and "improved" through human reason and technology.[20] Mechanistic and utilitarian views of the natural world further reinforced this separation, promoting the idea that nature was a resource to be exploited for human benefit rather than a system with which humans should cooperate.[21]

The rise of *capitalism* is often identified as a key driver of environmental domination. Under capitalist systems, nature became commodified, and the pursuit of continuous economic growth demanded ever-increasing resource extraction. Technological innovations amplified this process, enabling large-scale ecological manipulation and accelerating the pace of environmental degradation. The logic of capital accumulation prioritized short-term gains over long-term ecological stability, further entrenching a pattern of domination over natural systems.[22]

While these explanations provide valuable insights into the mechanisms of environmental degradation, they leave a fundamental question unanswered: Why has sociocultural niche construction—humanity's capacity to shape its environment—so frequently led to domination and exploitation rather than fostering more cooperative and sustainable relationships with nature?

Human societies possess advanced capacities for social learning, cooperation, and environmental management, which could, in theory, have been directed toward more sustainable interactions with the natural world. Yet, historical evidence suggests that these abilities were often channeled toward maximizing resource extraction and environmental control. Why did human societies not evolve stronger cultural norms of environmental stewardship, particularly as they gained more knowledge of ecological processes? Was this trajectory the product of contingent historical, economic, and technological pressures, or were there deeper cognitive or cultural reasons that led human societies down this path of domination?

This gap in the current understanding of sociocultural niche construction suggests a need for further inquiry. While the mechanisms of environmental domination are well documented, the reasons for the absence of alternative paths—those emphasizing mutualistic cooperation with nature—remain underexplored. Addressing these questions could provide essential insights into how human societies might shift toward more sustainable modes of interaction with the environment in the future.

An Alternative Explanation

Theories of resource maximization, competitive social structures, and dualistic worldviews offer partial explanations but fail to account for the broader, long-term dynamics of human–environment interactions. A potential reason for this limited interpretation may lie in the mismatch in time scales between sociocultural evolution and biological evolution.

Human societies have evolved rapidly in cultural and technological terms, while the natural systems they interact with evolve over much longer time scales. This discrepancy creates the illusion that humans are in control of Earth's future, as the fast pace of sociocultural innovation allows for unprecedented ecological transformation. However, this sense of control is often short-lived, as it overlooks the more complex processes of biological and ecological change that shape the biosphere over millennia. The rapid changes brought about by human sociocultural evolution may mask the long-term consequences of human actions on ecosystems, which continue to evolve on much slower time scales.

Earlier assumptions about the different time scales of ecological and evolutionary processes have shaped the unidirectional character of most empirical eco-evolutionary studies and can partly explain our lack of curiosity about the human role in shaping the evolutionary trajectory of planet Earth. However, recent evidence that significant evolutionary change does occur on a short time scale urgently challenges ecologists and evolutionary biologists to redefine the dynamic interplay between the two fields and to understand the interactions between human agency and eco-evolutionary feedback across different levels of biological organization. Humans are major drivers of micro-evolution. In human-dominated environments, selection pressures acting on traits can affect population dynamics by changing organisms' rates of survival or reproductive success, leaving a genetic signature that might affect community dynamics and ecosystem functions.

Moreover, current explanations are often constrained by a narrow historical time frame, focusing on the rise of agriculture, industrialization, and the impacts of human activity during the Anthropocene. This limited perspective obscures a deeper understanding of the role of humans in Earth's evolutionary history. While the Earth system is undergoing rapidly increasing transformation that coincides with the most recent Great Acceleration[23] of human activities, human transformation of Earth began much earlier.[24] As highlighted by scholars such as Matthew Edgeworth and Erle Ellis, this transformation should be understood as an event—materially unfolding through time—rather than constrained to a specific epoch.[25] If we place human niche construction within a broader eco-evolutionary context, it becomes possible to envision alternative paths—ones in which humans coevolve with their environments in more sustainable, mutualistic ways.

The current dominant narrative, which sees humans as separate from and in control of nature, may be a temporary phase in a much longer evolutionary story. By expanding time scales and conceptual lenses, it becomes clear that humans are not outside of nature but deeply embedded within Earth's ecosystems. Like all species, humans are subject to the long-term processes of adaptation, selection, and coevolution. Recognizing this interconnectedness offers the possibility of envisioning futures in which human societies evolve toward greater cooperation with other species and become integral parts of ecological systems rather than continuing cycles of domination and environmental degradation.

This broader perspective raises unasked questions: Might the apparent dominance of human societies over nature be an illusion, born from the time-scale mismatch between the rapid pace of sociocultural evolution and the slower rhythms of ecological and evolutionary change? If so, how might human societies realign with the long-term eco-evolutionary processes that govern the biosphere? Addressing these questions is crucial for understanding not only why human societies have followed a path of domination but also how they might forge an alternative path—one that emphasizes cooperation, adaptability, and mutualistic niche construction as key principles for thriving within Earth's dynamic systems.

Evolution of the Human Symbiotic Niche

Niche construction, as an evolutionary mechanism, underscores how organisms modify their environments and influence selection pressures on themselves and others. This process is fundamental to understanding how symbiotic relationships evolve and facilitate the coexistence of species. The concept of the *symbiotic niche*—where two or more species mutually shape each other's evolutionary paths—extends the idea of niche construction beyond competitive interactions. Symbiotic niches arise through mutualistic relationships, in which species benefit from each other's presence, expanding their ecological and evolutionary potential.

In mutualistic interactions, species co-construct their niches in ways that not only modify the environment but also alter evolutionary dynamics. Mycorrhizal fungi and plants provide a clear example: The fungi enhance nutrient and water absorption for the plants, while the plants supply carbohydrates to the fungi. This mutually beneficial relationship allows both species to thrive in environments that would otherwise be less favorable. The symbiotic niche they construct thus expands the ecological space in which they can survive, reproduce, and influence future evolutionary pathways.

The emergence of symbiotic niches represents a key shift in how we understand the interplay between ecological and evolutionary processes. While competitive interactions often limit niche space, mutualistic interactions expand it, creating new opportunities for species to adapt, evolve, and coexist. This expansion of niche space through cooperation highlights the importance of symbiotic relationships in driving evolutionary change. For example, species traits can evolve not just to cope with environmental pressures but also in response to the benefits derived from mutualistic partners. As these relationships evolve, they generate eco-evolutionary feedback loops, with changes in one species driving adaptations in another.

How do urban settlements change the mutualistic niche? And how does the emergent mutualistic niche alter symbiotic relationships among species? The answer to these questions depends on whether niche construction is considered a fundamental evolutionary process or simply an outcome. That is, do the selection pressures modified by environmental changes resulting from human niche construction influence the evolutionary trajectories of populations? It is widely acknowledged that niche construction is ubiquitous among living organisms and plays a significant role in shaping ecosystems. Evolutionary biologists debate whether the activities organisms engage in when constructing their habitat have evolutionary consequences comparable to those of natural selection.

The debate over whether niche construction is a fundamental evolutionary process or simply an outcome has particular relevance to human-altered environments. Niche construction modifies selection pressures, creating feedback loops that influence the evolutionary trajectories of populations and the ecological interactions between species. These dynamics are especially pronounced in urban environments, where human activities reshape habitats in ways that enable the emergence of novel mutualistic niches. By altering the conditions under which species interact, urban landscapes can drive the evolution of symbiotic relationships, highlighting the interconnected roles of humans and other organisms in co-constructing adaptive ecosystems.

A shared multispecies niche refers to the set of environmental conditions, interactions, and resources that enable the sustained coexistence of multiple species across their life cycles—across daily, seasonal, and generational cycles for all life stages. Grounded in niche theory, this perspective accounts for relational dynamics, ecological complexity, species autonomy, and adaptability over time and space. Each scale supports distinct functions, and species co-shape ecological processes across scales, from microhabitats to the city scale. In urban environments, these niches emerge through interactions among species and with human infrastructure, shaping and reshaping the conditions that support survival, reproduction, and ecological function. Designing for multispecies niches requires attention to how species use space, respond to environmental variability, and participate in ecological processes that sustain urban life.

Urban ecosystems serve as laboratories for redefining human–nature relationships. The challenge is to harness cities' transformative potential for the mutual benefit of all species. By integrating green infrastructure, promoting biodiversity, and designing spaces that accommodate both human and nonhuman needs, cities can become hubs of ecological innovation. Rethinking the interplay between human sociocultural evolution and eco-evolutionary dynamics offers fresh insights into aligning human activities with ecological processes, paving the way for a future in which human societies coexist with other organisms.

Cities as Hubs of Co-Constructed Niches

Understanding the evolution of cities as complex, multispecies environments reflects a deeper understanding of urban ecosystems as sites of co-constructed niches, where human and nonhuman species both shape and are shaped by the environment. In contrast to the current view of cities as human-dominated spaces, this new perspective highlights the roles of diverse organisms—plants, animals, insects, and microbes—in actively shaping urban ecosystems.

One fundamental aspect of cities as co-constructed niches is the emergence of mutualistic relationships between humans and other species. While in contemporary cities these relationships are typically limited to domesticated species integral to urban life, in multispecies cities, new symbiotic relationships with wildlife emerge through intentional human cooperation with other species, unfolding in collaborative ways rather than through top-down interventions, reflecting a shifting understanding of humans' place in nature.

In a bio-cooperative city of the future, urban ecosystems are shaped through coupled human–environment interactions. For instance, extensive networks of green corridors planted with native flowering species support pollinator populations such as bees and butterflies. These pollinator-friendly pathways, which include rooftop gardens, vertical farms, and balcony planters, enable pollinators to traverse the city freely.

The increased pollination services enhance urban agriculture and biodiversity in a mutualistic relationship that benefits both humans and pollinators.

Buildings evolve into habitats for various species through the adoption of bio-receptive materials and architectural features that encourage the presence of beneficial organisms. Walls constructed from mycelium-based composites not only provide structural insulation but also host beneficial microbes and insects. These organisms aid in air purification and pollutant degradation, enhancing indoor environmental quality while fostering habitats for other species. This symbiotic use of building materials exemplifies how architectural practices can intentionally support ecological processes. Architects and builders, motivated by sustainable practices, choose these materials to enhance building performance while simultaneously creating niches for other species.

Aquaponic systems are integrated into public spaces and residential areas, creating closed-loop ecosystems where fish and plants coexist symbiotically. The waste produced by the fish supplies essential nutrients for the plants, which in turn filter the water, maintaining a healthy environment for the aquatic life. These systems supply residents with fresh produce and protein sources while establishing habitats for various aquatic organisms, demonstrating a sustainable approach to food production that aligns with natural ecological cycles.

Urban architecture accommodates wildlife through design features that support species such as birds and bats. High-rise buildings incorporate nesting sites for swifts and swallows, with specialized ledges and cavities that facilitate nesting without compromising structural integrity. These birds assist in controlling insect populations, providing a natural form of pest management. Similarly, the city includes bat bridges and tunnels connecting roosting sites, supporting nocturnal wildlife that contributes to pest control. Soft, wildlife-friendly lighting minimizes disruption to natural behaviors, indicating a thoughtful integration of human and wildlife needs.

Urban green spaces become rich ecosystems as communities engage in planting fruit trees, nut trees, and edible perennials. These plants offer food resources for both humans and wildlife, including birds, insects, and small mammals. Community gardens and edible landscapes emerge from collective efforts by residents who aim to create sustainable and nourishing environments. These shared spaces facilitate interactions among species and promote nutrient cycling through natural processes like seed dispersal and decomposition, exemplifying intentional cooperation without centralized planning. Community harvest events promote shared use of these resources, fostering a sense of community and stewardship over local ecosystems. This approach enhances urban food security while supporting biodiversity.

To support beneficial insect populations, features like insect hotels are installed in parks and green spaces. These structures provide habitats for species such as ladybugs, bees, and beetles, which contribute to pollination and natural pest control. Reducing reliance on chemical pesticides aligns with sustainable urban management practices and promotes ecological balance. Safe movement for larger wildlife species is realized through the construction of urban wildlife crossings such as overpasses and underpasses. These crossings are landscaped to resemble natural habitats, allowing species like deer and foxes to navigate the urban landscape without risk of vehicle collisions. This connectivity promotes genetic diversity and ecological resilience within urban wildlife populations.

Water management infrastructure incorporates microbial biofilters, such as constructed wetlands populated with specific microbes and plants that naturally degrade contaminants. These systems purify wastewater before it reenters the environment and create habitats for amphibians and insects. Wetland areas are restored or created through grassroots environmental projects, enhancing natural water purification processes and providing habitats for diverse microorganisms, plants, and animals.

Organic waste management is addressed through community composting centers that double as worm farms. Earthworms process organic waste into nutrient-rich fertilizer for use in community gardens, closing the nutrient loop. Additionally, the worms serve as a food source for birds and small mammals, integrating waste management with food webs and supporting local biodiversity.

Smart technologies represent a convergence of biology and urban infrastructure design. Innovative approaches to urban lighting include the use of bioluminescent plants and algae to illuminate streets and parks. This reduces energy consumption and minimizes light pollution, benefiting nocturnal wildlife and contributing to more natural nighttime environments. Adaptive soundscapes employ sound technologies that replicate natural environments. Amplifying sounds like birdsong in specific areas creates calming atmospheres, which have been shown to reduce stress in humans and attract wildlife. This sensory enhancement of urban spaces contributes to human well-being and supports ecological diversity. Interactive digital ecosystems provide residents with real-time data on local wildlife populations and environmental conditions through public displays. These systems enhance awareness of urban ecology, encouraging community engagement and fostering a sense of connection to the shared environment. This technological integration supports environmental education and promotes stewardship.

In bio-cooperative cities, the emergence of mutualistic relationships and co-constructed niches is an outcome of humans actively engaging with their environment in cooperative ways. The urban ecosystem evolves as a complex adaptive system in which individual and collective actions contribute to sustainability and resilience. This paradigm represents a shift from anthropocentric urban development to a model where humans participate as integral members of a multispecies community, fostering the well-being of all species through multispecies cooperation.

These mutualistic interactions illustrate the ways in which cities can function as ecosystems that support both human and nonhuman life. Cities, far from being devoid of nature, are alive with species that adapt to and co-construct their environments alongside humans. This process of co-construction extends beyond individual species to the entire urban ecosystem, influencing energy flows, material cycles, and evolutionary dynamics.

The resilience of bio-cooperative cities emerges from dynamic interplay between species and their urban environments. As integral components of green and blue infrastructure, organisms actively shape urban ecosystems through their ecological roles and interactions. For example, trees, shrubs, grasses, and aquatic plants regulate carbon, water, and temperature, driving essential ecosystem functions (table 6.1). However, their ability to sustain these roles depends on the urban environment supporting their evolving niches—including water availability, soil conditions, climate, and species interactions (table 6.2).

Table 6.1

Ecosystem functions supporting climate change adaptation

Ecosystem function	Heatwaves	Floods	Droughts	Wildfires
Carbon sequestration	Urban forests and green roofs absorb CO_2, reducing greenhouse gases that contribute to heatwaves.	Wetlands, trees, and soil carbon storage help mitigate floods by absorbing CO_2.	Forests and grasslands capture carbon, indirectly mitigating drought by lessening climate change severity.	Forests and green spaces store carbon, reducing atmospheric CO_2 and mitigating fire risks.
Water regulation	Green infrastructure helps manage stormwater, reducing surface water that heats up urban areas.	Wetlands, rivers, and green spaces absorb and slow water flow, reducing peak flood impacts.	Wetlands and permeable landscapes store water, maintaining levels during dry periods.	Riparian zones and water bodies act as natural firebreaks, helping regulate water availability near wildfires.
Temperature regulation	Trees and green spaces provide shade and evapotranspiration, reducing urban heat island effects.	Green spaces and trees in flood-prone areas provide shade, cool flooded areas, and prevent stagnant heat.	Trees and vegetation offer shade and humidity, maintaining cooler microclimates during drought conditions.	Vegetation provides shade, cooling, and fuel breaks, reducing fire intensity and spread.
Soil stability	Green spaces and permeable soils absorb and cool runoff, limiting heat accumulation in cities.	Root systems stabilize soils and prevent erosion during floods, keeping infrastructure resilient.	Stable soils with vegetation retain water—critical during droughts for preventing desertification.	Root systems prevent soil erosion after fires, reducing the impact of post-fire floods and landslides.
Biodiversity support	Habitats for species that contribute to pollination ensure the resilience of urban food supplies during heat.	Biodiverse wetland systems support aquatic life, mitigating the spread of flood-related pollutants.	Drought-resistant species provide food and water resources for species during dry periods.	Diverse species provide habitats that recover faster after fires, aiding ecosystem resilience.

Table 6.2
Species supporting flood mitigation

Ecosystem function	Trees	Shrubs	Grasses	Aquatic Plants
Carbon sequestration	Wetland trees such as mangroves capture CO_2, helping reduce flood-related emissions.	Shrubs such as willows stabilize carbon in soil, reducing soil erosion.	Grasses such as reed grass capture and store carbon in floodplains.	Aquatic plants such as seagrasses store carbon in submerged soils, reducing flood emissions.
Water regulation	Trees such as bald cypress slow down water flow, mitigating flood peaks.	Shrubs along riverbanks stabilize banks and slow water runoff, reducing flood intensity.	Floodplain grasses absorb and hold water, reducing peak flows.	Aquatic plants trap sediment and slow water flow, aiding flood control.
Temperature regulation	Tree canopies cool floodwaters, reducing thermal pollution in waterways.	Shrubs near water bodies provide cooling shade to maintain habitat temperature.	Grasses in wetlands reduce surface heating, keeping floodwaters cooler.	Aquatic plants maintain cool water temperatures, beneficial during flooding.
Biodiversity support	Wetland trees provide habitats for waterfowl and fish species.	Shrubs support aquatic insects, providing food for fish and birds.	Floodplain grasses are habitats for wetland animals, supporting biodiversity.	Aquatic plants are essential for fish nurseries, contributing to aquatic biodiversity.

Flood mitigation illustrates how species and their environments coadapt to sustain ecosystem functions that support adaptation to climate change and enhance resilience (figure 6.1). Mangroves buffer coastal areas from storm surges, but only where tidal flows nourish their root systems. Willows stabilize riverbanks, yet their capacity to prevent erosion depends on consistent moisture levels. Seagrasses absorb wave energy and trap sediments, but their effectiveness relies on water clarity and stable seabed conditions (table 6.3). These relationships illustrate the fundamental principle of co-constructed niches: Ecosystem functions emerge from the interactions between species and their urban environments rather than from species' presence alone.

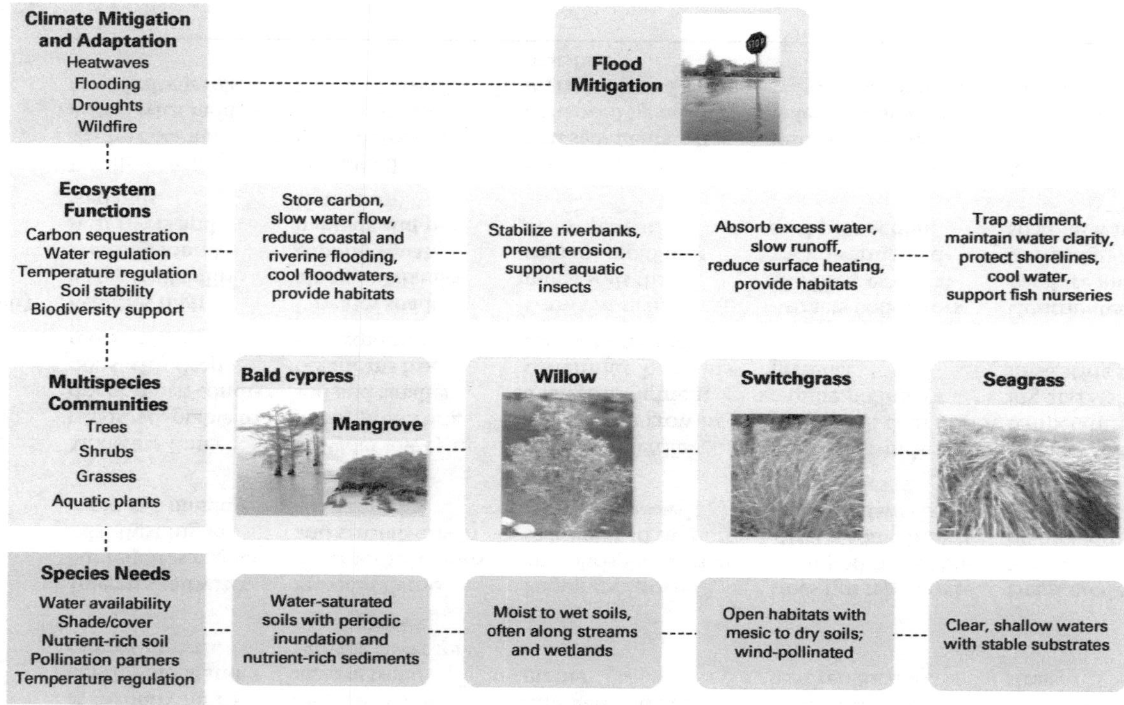

Figure 6.1

Multispecies climate mitigation and adaptation. *Source*: Author-created image. Photographs: Flood, Wikimedia, Roc0ast3r, public domain (CC0). Bald cypress (*Taxodium distichum*), iNaturalist, Craig Martin, public domain (CC0). Mangrove (*Rhizophora mangle*), iNaturalist, Zoltán Stekkelpak, public domain (CC0). Willow (*Salix* spp.), iNaturalist, 葉子, public domain (CC0). Seagrass (*Posidonia oceanica*), iNaturalist, Raffaele Maiorano, public domain (CC0). Switch-grass (*Panicum virgatum*), iNaturalist, peakaytea, public domain (CC0).

Table 6.3

Species needs and requirements for flood mitigation

Species	Water availability	Shade/cover	Nutrient-rich soil	Pollination partners	Temperature regulation
Oaks (*Quercus*)	Requires moderate to high water availability for root development and canopy growth.	Needs canopy shade for young saplings, while mature trees provide cover for other species.	Thrives in nutrient-rich soils with ample nitrogen and organic matter.	Supported by pollinators like bees and moths essential for seed production.	Can regulate local temperatures by providing shade, reducing heat stress.
Mangroves	Tolerates saltwater but requires regular tidal flushing to avoid salt buildup.	Provides dense cover for fish, birds, and crustaceans in coastal zones.	Requires nitrogen and phosphorus-rich sediments to support rapid root growth.	Does not rely heavily on pollinators but attracts some insects for secondary ecological roles.	Helps moderate temperatures in coastal areas by providing shade and cooling waters.
Switchgrass (*Panicum*)	Moderate water needs, adaptable to dry or moist conditions with deep roots.	Provides low-lying cover, protecting soil and smaller organisms from sun exposure.	Thrives in fertile soil but can grow in a range of nutrient conditions; helps in soil carbon storage.	Primarily wind-pollinated; does not require pollinator presence.	Reduces soil surface temperatures by covering and stabilizing soil, mitigating heat.
Willows (*Salix*)	Requires high water availability; typically found near rivers and wetlands.	Provides shade along riverbanks, cooling aquatic habitats and protecting soil.	Grows in rich, moist soils that support rapid root development.	Attracts pollinators like bees, essential for flowering and seed production.	Contributes to temperature moderation in riparian zones by shading water.
Seagrasses	Requires clear, shallow coastal waters with moderate salinity levels.	Provides dense underwater cover for fish, shellfish, and juvenile species.	Requires nutrient-rich sediments, often enhanced by nearby mangroves and marshes.	Does not require terrestrial pollinators; relies on water for pollen distribution.	Moderates temperatures in coastal habitats, providing cooler microclimates underwater.

Mainstream urban design often treats natural elements as isolated objects, emphasizing their spatial configuration while overlooking the ecological processes that sustain biodiversity and resilience. By focusing on the placement and distribution of green spaces, water bodies, and other environmental features as static elements, this approach risks neglecting the dynamic interactions that define and maintain urban ecosystems. A more integrated perspective recognizes that ecological resilience depends not only on the arrangement of habitats but also on the functional relationships among species and their evolving niches.

By designing with species' needs in mind, cities can move beyond a static, habitat-based approach to one that fosters ecological processes, enabling biodiversity to thrive and strengthening climate adaptation. A socioecological, traits-based niche perspective further enhances our understanding of how species characteristics and behaviors shape urban adaptation and coexistence, emphasizing the importance of designing for multispecies interactions rather than merely setting aside space for nature.

Given the profound influence of cities on global ecosystems, there is a compelling opportunity to reimagine urban areas as sites for cooperative, evolving niche construction. The concentration of social, technological, and economic innovations within cities could be leveraged to create urban systems that support biodiversity, mitigate environmental degradation, and enhance resilience in the face of climate change.

7 Eco-Evolutionary Dynamics

Imagine a city where infrastructure adapts dynamically to environmental change. Networks of green corridors and blue spaces shift in response to climatic fluctuations, urban rhythms, and multispecies interactions. Waterways and vegetated zones expand and contract with seasonal changes, managing floods while enabling species movement. Buildings incorporate living systems that modulate temperature, filter air, and evolve to support urban biodiversity. Sensors distributed across the city monitor soil moisture, air quality, and species populations, triggering timely, localized adjustments to sustain ecological functions. Infrastructure and species coevolve: As plants, animals, and microbes adapt to urban conditions, they reshape the systems that support them. The city operates as an integrated, evolving system, attuned to both short-term stressors and long-term eco-evolutionary dynamics.

It's 2080 and several cities around the globe have emerged as models of how infrastructure can evolve symbiotically with the natural environment. Accelerating climate change, combined with a deeper understanding of eco-evolutionary dynamics, has prompted a shift from rigid static systems to adaptive living systems that coevolve with their environments. At the Massachusetts Institute of Technology, a graduate planning studio gathers to present their findings on how this transformation unfolded. Five teams have traveled to different cities and studied how infrastructure became an active participant in ecological and evolutionary processes shaping urban ecosystems.

The first team focused on Ho Chi Minh City, where the infrastructure evolved to embrace, rather than resist, natural cycles of flooding.

"Standing by the Saigon River," one student began, "the mayor explained how the city stopped fighting the floods and learned to live with them."

Streets were redesigned with porous materials to absorb monsoon rains and release water gradually, alleviating pressure on drainage systems. The mayor had said, "They no longer act as barriers—they breathe with the rains."

Riverbanks, once hardened against water, were reimagined as adaptive floodplains. These floodplains, dynamically expanding and contracting with the river's ebb and flow, were seamlessly integrated with the Can Gio mangrove ecosystem. The result: a network of natural habitats that buffer against heavy rains while enhancing ecological functions, turning flood management into an opportunity for urban resilience and biodiversity.

Tokyo's blue-green corridors were highlighted by the second team as exemplary urban infrastructure fostering species adaptation. "Vegetation in Tokyo has evolved to withstand rising urban temperatures and increased pollution," a student explained. "By expanding forest cover and selecting plant species capable of tolerating high nitrogen deposition and drought, these ecosystems now thrive where others falter."

One such species, the Japanese zelkova (*Zelkova serrata*), has adapted to endure high nitrogen levels, providing critical habitat for urban wildlife. The tree supports insects and birds that have also adjusted to the altered environment. This strategic planting not only enhanced urban ecosystem resilience but also created a feedback loop: As these trees and plants efficiently filter pollutants, they simultaneously form vital corridors for species whose movement patterns have adapted to the seasonal availability of food and shelter within the city's growing green networks.

In New York City, the third team focused on living shorelines. "The oyster reefs here," a student explained, "have been deliberately cultivated to evolve in response to rising sea levels and increasingly acidic waters." City planners described how selective breeding programs encouraged the survival of oysters with greater tolerance to these conditions. Over time, these resilient oysters strengthened the reefs, enhancing their ability to absorb storm surge energy and protect the coastline. "These reefs," another student noted, "aren't just reducing wave energy—they're evolving ecosystems. Infrastructure, both physical and biological, now cleans the water, provides habitat for marine species, and adapts to environmental changes. As the oysters grow more resilient, the shoreline itself becomes increasingly effective at safeguarding the city."

The fourth team explored Singapore, focusing on its modular green roofs, which exemplify eco-evolutionary interactions at a different level.

"The engineers we spoke with explained that the plants on these roofs are no longer the same as they were a decade ago," a student said. "Over time, species composition has shifted, with drought-tolerant plants becoming dominant through differential survival and persistence in response to the urban heat island effect." This shift has enhanced the roofs' water retention efficiency during monsoons and their cooling capacity during dry spells. "Plants better suited to these harsh urban conditions thrive, and their evolution directly influences how effectively the infrastructure performs. It's an ongoing evolutionary process."

The fifth team, visiting Dhaka in Bangladesh, observed how the city's floodplains have coevolved with its river systems and the species inhabiting them. "A chief planner explained that certain fish species, for instance, have adapted their breeding cycles to align with the periodic expansion and contraction of the floodplains," a student shared. "In turn, the presence of these species has enhanced the floodplains' ecological functioning, including nutrient cycling and water purification." Designed to manage seasonal flooding, the floodplains have become an integral part of the city's ecological fabric. The infrastructure and the species have evolved together, creating a more resilient system.

At the conclusion of the presentations, the instructor addressed the students. "The ideas you've presented today indicate a fundamental change in our understanding of urban infrastructure. These systems are no longer static structures designed to withstand environmental forces, as they have been historically. Rather, they are dynamic, living systems that evolve alongside the ecosystems they inhabit."

After a short pause, she said, "What's evident is that the role of eco-evolutionary dynamics is crucial in shaping these systems. Infrastructure reshapes the urban environment influencing the species that inhabit it—plants, animals, and microorganisms—and their evolution determines how well infrastructure functions and changes over time. These eco-evolutionary processes form the basis of future resilient cities."

The instructor then posed a question: "What enabled this shift? It wasn't just technology or policy changes. It was the growing urgency of climate change, the advancement of eco-evolutionary science, and, perhaps most importantly, a new understanding of our relationship with the natural world—not as its controllers but as participants in its evolutionary processes."

She concluded, "This shift in mindset has made possible the innovations we see today. In these cities, we no longer resist nature—we evolve with it. The real challenge, moving forward, is not just to replicate these innovations but to understand that we humans are part of the natural world, not separate from it."

Designing for Evolutionary Potential

Building resilient cities in a rapidly changing world requires a shift in perspective—one that embraces the dynamic interplay of ecological and evolutionary processes. Resilience cannot be achieved through strategies that assume environmental stability and seek to preserve nature unchanged. Instead, it must support systems that adapt, evolve, and co-create with shifting urban ecosystems. This reframes urban planning from preserving current conditions to enabling continual adaptation and coevolution.

Resilience hinges on maintaining evolutionary potential—the ability of species and ecosystems to adapt to dynamic conditions.[1] Evolution underpins life's capacity to respond to change, shaping traits and interactions that enhance survival. Preserving genetic and phenotypic diversity within urban ecosystems increases the likelihood of species thriving amid environmental shifts. This diversity sustains critical ecosystem functions, including pollination, nutrient cycling, and climate regulation, essential for urban sustainability.

This perspective demands rethinking urban planning strategies. The assumption of a fixed "target condition" gives way to a planning framework in which goals are continually revised in response to environmental change. Planners must anticipate the cascading effects of eco-evolutionary processes, from changes in species traits to feedback loops that redefine ecosystem interactions. Urban systems become more complex and less predictable, but this very uncertainty fosters opportunities to co-create hybrid systems that leverage nature's inherent adaptive capacity.

This evolutionary lens is no longer theoretical. It is beginning to inform the design and management of urban infrastructure. Nature-based solutions—interventions that work with natural processes to protect or restore ecosystems—offer a pathway toward realizing this shift. These systems have the potential to adapt over time. For instance, living shorelines adapt as oysters respond to rising sea levels and increasing acidity, enhancing both coastal protection and biodiversity. Similarly, green roofs support plant communities that adjust to urban heat and drought, improving their cooling and stormwater management capacities.[2]

The strength of these solutions lies in feedback loops between ecological processes and human systems. Urban pressures, like heat and pollution, drive species-level adaptations, such as drought-tolerant plants or microbes that metabolize contaminants in constructed wetlands. These adaptations enhance the long-term performance of nature-based solutions, creating infrastructure that becomes more effective. However, these dynamics also introduce evolutionary trade-offs. For instance, species adapting to heat stress may lose other functions, such as pollination or carbon sequestration. Anticipating and managing such trade-offs is critical to sustaining the long-term efficacy of nature-based solutions.[3]

Evolutionary changes in species traits ripple through urban ecosystems, influencing their functionality. Adaptations—such as deeper-rooted plants increasing soil moisture retention or microbes evolving pollutant tolerance—can trigger feedback loops that may strengthen or erode resilience. Understanding these dynamics allows planners to design systems that adapt and thrive under rapidly changing conditions while explicitly accounting for trade-offs between short- and long-term ecosystem benefits.

Nature as Evolving Infrastructure

Nature-based solutions are increasingly recognized as essential components of resilient urban infrastructure. These systems—green roofs, bioswales, constructed wetlands, and living shorelines—are designed to mitigate climate impacts, enhance ecosystem functions, and support biodiversity. Bioswales, for instance, filter stormwater and remove pollutants, reducing flood risks, while constructed wetlands purify water and promote habitat diversity. Living shorelines buffer coastal areas from storm surges, reduce erosion, and trap sediments.[4]

The potential of nature-based solutions extends beyond immediate ecological functions. Urban ecosystems are inherently dynamic, shaped by ecological interactions and evolutionary processes. As species adapt to urban conditions—altered temperatures, pollution, and habitat fragmentation—their evolutionary trajectories affect their performance. For example, green roofs depend on plant species that might evolve traits such as improved drought tolerance or water-use efficiency, directly influencing their ability to cool urban areas and manage stormwater.

Nature-based solutions represent a critical asset in preserving a city's evolutionary potential. By integrating eco-evolutionary feedback mechanisms into their design, we can enhance the effectiveness of these solutions in mitigating the impacts of climate change.

Evolutionary Trade-Offs and the Functionality of Nature-Based Solutions

One of the central mechanisms through which eco-evolutionary dynamics influence the functionality of nature-based solutions is the emergence of evolutionary trade-offs. Species that evolve to survive in highly modified urban environments often develop traits that enable them to cope with new pressures. However, these adaptations can come at the cost of other important functions. For example, plants on green roofs may evolve to prioritize rapid growth in response to limited soil depth and water availability. Although this adaptation enables short-term survival, it can result in reduced root development, which decreases water retention and weakens the long-term stability of the green roof system.

Similarly, urban heat islands create selective pressures that drive rapid evolutionary changes in thermal tolerance. Pollinators, such as bees, may evolve to withstand higher temperatures; however, this adaptation could lead to reduced biodiversity, as only heat-tolerant species would survive. Such changes could disrupt pollination dynamics, potentially reducing the resilience of plant species that depend on diverse pollinators. Furthermore, species that evolve to thrive in urban settings might become less competitive in natural ecosystems, narrowing the gene pool and potentially reducing genetic diversity.

Another trade-off can be seen in the use of predators for pest control in urban environments. Predators such as ladybugs or parasitic wasps are often introduced to manage pest populations. However, over time, prey species may evolve resistance to these predators, creating imbalances in predator–prey dynamics. This can lead to pest outbreaks that are no longer effectively controlled by natural predators, destabilizing the ecosystem. In such cases, evolutionary adaptations that provide short-term advantages to prey species can undermine the long-term stability of the ecosystem, underscoring the complex and often unpredictable nature of eco-evolutionary feedback in urban systems.

These trade-offs are not confined to a single type of ecosystem. They are a fundamental aspect of eco-evolutionary dynamics across urban environments. To ensure that nature-based solutions continue to function effectively, urban planners and ecologists must account for these evolutionary trade-offs. Incorporating eco-evolutionary dynamics into the design of nature-based solutions is essential to promoting ecological and evolutionary resilience.

Evolutionary Trade-Offs in Coastal Environments

Coastal cities face unique challenges related to climate change, such as rising sea levels, increased storm surges, and coastal erosion. Nature-based solutions, such as seagrass beds, salt marshes, and mangroves, are critical for protecting coastal infrastructure by buffering wave action, slowing water movement, and trapping sediment. These ecosystems not only provide physical protection but also deliver essential ecosystem functions, such as nutrient cycling, carbon sequestration, and habitat provision for both human and nonhuman communities.

However, the effectiveness of these coastal defenses is affected by eco-evolutionary changes driven by urbanization, pollution, and climate change. As species adapt to the selective pressures of urban coastal environments, their evolutionary responses can significantly affect the performance of these systems. While these adaptations might sometimes enhance functionality, they often introduce trade-offs that may compromise long-term resilience. These dynamics must be carefully considered when designing and implementing nature-based solutions to ensure their sustained success.

Seagrass meadows, such as those formed by *Zostera marina*, play a vital role in coastal protection by trapping sediment and improving water clarity. However, these systems are highly vulnerable to the effects of urbanization and climate change. Increased nutrient runoff, leading to eutrophication, has caused widespread die-offs of seagrass in many coastal regions. Even after nutrient levels are reduced, restored seagrass beds frequently struggle to adapt to fluctuating salinity and pollution levels that are exacerbated by climate change.

Seagrasses exposed to these stressors may undergo trait shifts such as reduced stature or altered root structures, which enable them to survive in polluted or high-salinity environments. While these adaptations may help seagrass populations persist in the short term, they often come at the cost of their ability to effectively trap sediment and buffer wave action. As a result, seagrass beds may lose their capacity to protect coastal infrastructure from flooding and erosion, underscoring the trade-offs that arise when evolutionary potential is not fully integrated into restoration efforts.

Salt marshes provide another example of how evolutionary dynamics can influence the success of coastal nature-based solutions. Salt-tolerant grasses, such as *Spartina* species, which play a key role in stabilizing sediment and reducing coastal erosion, may evolve to cope with rising salinity and other stressors associated with urban development and sea-level rise. However, these adaptations may affect their growth patterns, resulting in reduced biomass or less effective sediment stabilization. As species evolve to withstand harsh conditions, they may allocate resources toward survival, such as salt tolerance, with downstream effects on ecological functions including nutrient cycling and habitat provision for wildlife.

The Case of Chesapeake Bay

The Chesapeake Bay restoration provides a compelling real-world example of how evolutionary resilience and ecological dynamics intersect within the framework of nature-based solutions. While seagrass systems and salt marshes undergo gradual evolutionary changes, the Chesapeake Bay case illustrates how species-level resilience, exemplified by oysters, can enhance broader ecosystem restoration efforts and the effectiveness of these approaches.[5]

Chesapeake Bay has experienced extensive restoration efforts to reverse degradation caused by overfishing, pollution, and habitat loss. One of the most notable successes has been the resurgence of oyster populations. In 2014, the Army Corps of Engineers deposited fossilized oyster shells into Virginia's tributaries, including Baines Creek, to restore oyster reefs. By 2020, hundreds of oysters per square meter were thriving on these reefs, with some reaching ages previously unattainable due to disease outbreaks introduced into the bay.

This restoration effort highlights evolutionary resilience: the surviving oyster populations likely adapted to withstand diseases that once devastated them. Unlike the slower adaptations seen in seagrasses and salt marshes, the rapid recovery of oysters illustrates how nature-based solutions can leverage species' adaptive capacity to maintain long-term resilience. Acting as ecosystem engineers, oysters filter water, improve water quality, and support the recovery of seagrasses and other species. This positive feedback loop between oysters and their environment strengthens the resilience of the entire ecosystem, demonstrating the crucial role of population-level adaptation in the success of these approaches.

Despite these successes, the Chesapeake Bay restoration continues to face significant challenges. Nutrient pollution from agricultural runoff—specifically nitrogen and phosphorus—remains a persistent issue, exacerbating algal blooms and creating dead zones in tributaries such as the Choptank River. These nutrient-driven stressors demonstrate how evolving environmental conditions, such as nutrient pollution and climate change, can undermine restoration efforts if long-term ecological adaptability is not adequately addressed.

Harnessing the Evolutionary Potential of Microbes

Urban wetlands often host microbial communities that play a key role in detoxifying water. These microbes may evolve in response to pollutants, such as industrial runoff, developing traits that enable them to break down harmful chemicals more efficiently.[6] As these microbial communities adapt, the capacity of wetlands to filter contaminants increases, making wetlands a robust solution for urban water management.[7]

Microbial communities are foundational to many ecosystem functions, especially in urban environments where water quality, soil health, and pollution management are critical. Often overlooked in discussions of urban resilience and nature-based solutions, microorganisms play a vital role in maintaining urban ecosystem health. These organisms are highly dynamic, evolving rapidly in response to environmental pressures, which creates both opportunities and risks: the potential of microbes to mitigate pollution and the possibility of antibiotic-resistant bacteria from wastewater treatment plants accidentally getting into natural bodies of water.

Microbial communities possess remarkable adaptability, making them key agents in bioremediation. In green infrastructure—such as bioswales, constructed wetlands, and permeable pavements—microbes break down pollutants like hydrocarbons, heavy metals, and pharmaceuticals, helping to purify urban water supplies. As microbial communities are exposed to various pollutants, they evolve and develop traits that enhance their capacity to metabolize and detoxify harmful substances.

Bioremediation in Constructed Wetlands

In China's Pearl River Delta, constructed wetlands are used to treat industrial wastewater contaminated with textile manufacturing pollutants, including dyes and aromatic hydrocarbons. Over time, microbial communities in these wetlands—particularly bacterial strains such as *Pseudomonas*—have evolved to better degrade these complex pollutants. Bacteria have developed specific enzymes that allow them to break down hydrocarbons more efficiently, improve water quality, and support the overall health of the ecosystem.[8]

As bacteria evolved to manage increasingly toxic compounds, wetlands became more effective at filtering water, creating a positive feedback loop between microbial evolution and environmental health. This ongoing adaptation is essential for maintaining the resilience of nature-based solutions in the face of fluctuating pollution levels. However, this dynamic process carries several risks. Microbial communities that evolve to specialize in metabolizing specific pollutants may lose versatility and become less capable of adapting to new environmental conditions or contaminants. Therefore, managing the evolutionary dynamics of microbial communities is crucial to ensuring the long-term success of bioremediation-based solutions.

The Threat of Antibiotic Resistance

While microbial evolution can be harnessed for positive outcomes, such as bioremediation, it also poses significant challenges, particularly the evolution of antibiotic resistance. Urban environments—in which human activities drive the extensive use of antibiotics in health care, agriculture, and animal husbandry—have become hotspots for the rapid evolution and spread of antibiotic-resistant bacteria.[9] This is not only a growing public health concern but also a significant ecological challenge for bio-cooperative cities.

Antibiotic resistance occurs when bacteria evolve mechanisms to survive exposure to drugs intended to kill them. The problem is exacerbated in urban water systems, where untreated or poorly treated wastewater can carry antibiotics, pharmaceutical residues, and resistant bacteria into natural water bodies. Wastewater treatment plants, integral to urban infrastructure, may inadvertently become reservoirs for resistant bacteria, allowing these microbes to spread into the broader environment.

For example, studies have shown that wastewater treatment plants near major cities in India and China are hotspots for antibiotic-resistant bacteria, such as strains of *Escherichia coli* and *Klebsiella pneumoniae*. These bacteria have evolved resistance to multiple classes of antibiotics, including penicillin, tetracycline, and cephalosporins, making them difficult to treat with conventional medical therapies. Once released into natural water systems, these resistant strains can spread to other microbial communities, wildlife, and humans, creating an ecological and public health crisis.

The evolution of antibiotic resistance exemplifies eco-evolutionary feedbacks, where human activities (widespread antibiotic use) create selective pressures that drive the evolution of resistance traits in microbial populations. In turn, these resistant bacteria spread through urban environments, particularly in densely populated areas with inadequate sanitation and water treatment facilities. The movement of resistant bacteria through water systems underscores the interconnectedness of urban ecosystems and the global scale of this evolutionary challenge.

Managing Evolutionary Feedback

Addressing the threat of antibiotic resistance requires a dual approach, combining technological advancements and ecological management. On one hand, urban infrastructure must be upgraded to improve wastewater treatment and reduce the release of pharmaceutical pollutants. On the other hand, cities need to closely monitor microbial evolution, utilizing genomic tools to track the spread of resistance genes through water systems and urban environments.[10]

Solutions might include integrating constructed wetlands designed not only for pollution mitigation but also for reducing antibiotic-resistant bacteria by fostering microbial communities that can outcompete or neutralize resistant strains. Additionally, reducing the overuse of antibiotics in health care and agriculture, along with more stringent sanitation and water treatment policies, can help slow the spread of resistance.

Proactively managing microbial communities within urban infrastructure is essential to harnessing their evolutionary potential for positive outcomes while minimizing risks such as antibiotic resistance. By understanding and guiding microbial evolution, cities can create dynamic systems that adapt to and evolve with environmental pressures, ensuring that urban ecosystems remain resilient in the face of both natural and human-induced challenges.

In the imaginary seminar room at MIT in 2080, full of sketches and blueprints of evolving cities, the idea of cities embracing evolution is not a futuristic vision; it is a reality. The many types of infrastructure that the students studied were designed to adapt and evolve alongside the ecosystems they inhabited. Urban resilience has shifted to a cooperative approach, where human systems and nature work together, where cities and their environments coevolve.

The instructor's message was clear: "The key is working with nature, recognizing the agency of other species, and building systems that evolve together."

Today, eco-evolutionary dynamics offer a foundation for creating adaptive and resilient cities. By recognizing the evolutionary capacity of organisms, cities can develop bio-cooperative systems in which infrastructure and natural processes function in an interdependent manner. This approach fosters infrastructure systems that evolve with species, adapting to environmental changes as part of a living, dynamic ecosystem.

Incorporating evolutionary principles into nature-based solutions helps cities create adaptive systems that respond to ecological processes. These infrastructure systems evolve with the changing environment, ensuring that cities thrive under dynamic conditions.

This method builds resilience and ensures long-term sustainability. The future of cities lies in co-creating nature and designing systems that evolve within the living world. By adopting a multispecies approach and integrating eco-evolutionary feedback into urban planning, cities can become dynamic, adaptive systems, fully integrated with their environments.

8 Multispecies Equity

In the courtroom of Sōma, a small Japanese city on the east coast, north of Tokyo, after an extraordinary hearing that captivated the entire city, something unprecedented took place: Justice was redefined. The plaintiffs in this case were not people, but the city's trees, birds, and insects—species that are fundamental to the urban landscape yet often overlooked. Their claim was not for survival, but for recognition. They stood in defense of their right to be seen as vital members of the city's ecosystem, demanding protection and respect, as their existence is inseparable from the city's life.

The case did not follow the usual legal procedures. Instead of human plaintiffs pleading for their interests, the court had to grapple with a groundbreaking question: Could nonhuman species—creatures who add to the city's life support systems—be considered citizens? Could they claim the "right to the city" just as humans do? The argument was simple yet profound: If these species contribute to the environment that makes the city livable, shouldn't they be acknowledged as citizens too, with rights to exist and thrive?

This imaginary scenario is part of a speculative, immersive exercise designed to explore what multispecies justice could look like in an urban context. In this constructed setting, Sōma becomes a site where planners, scholars, and citizens engage in a future in which the rights and recognition of nonhuman entities are considered central to urban life. "Multispecies participatory futures" are collective imaginary practices that help explore potential futures.[1] By stepping into these hypothetical realities, participants can reconsider how cities might function if they were to truly embrace the interconnectedness of all species, challenging the boundaries of justice, citizenship, and governance in the complex ecosystems that define urban environments.[2]

The idea of extending the "right to the city" to species other than humans is rooted in earlier concepts of nonhuman entanglement and has only recently emerged in urban planning.[3] Posthumanist approaches integrate the needs of nonhuman species—animals, plants, and even landscapes—into city design and governance.[4] These frameworks recognize the value of nonhuman actors, viewing them as active participants in the democratic process of urban life, and emphasize the need to reframe justice as something that applies to all urban inhabitants. How can the "right to the city" for nonhuman species become reality?

The imaginary courtroom in Sōma became the stage for a fundamental shift in thinking. This isn't a scene from a futuristic novel; it reflects a real, growing global movement. Around the world, legal frameworks are beginning to recognize the rights of rivers, forests, and ecosystems, breaking with the belief that justice applies only to humans. But in cities—where the majority of us live—we still govern as though nonhuman species are just part of the background, their presence incidental.

Sōma's imagined scenario offers a vision of what it might mean for a city to embrace a broader conception of justice through a new legal framework. This approach reshapes human identity by recognizing our interdependence with nonhuman entities and acknowledging their value and agency, incorporating them as active participants in urban democracy and planning.[5]

Envisioning a Future of Multispecies Justice

As we venture further into the possibilities of multispecies justice, we confront big questions: Who exactly falls under the umbrella of justice? Does it extend to sentient animals, all living organisms, or even entire ecosystems? Expanding the boundaries of justice to include nonhumans raises intricate challenges, especially when different justice claims come into conflict. Traditional theories often limit rights and responsibilities to humans, overlooking the countless contributions of nonhuman entities that sustain our world.

Emerging perspectives, such as sentientist justice, argue for extending justice to sentient nonhuman animals based on their capacity for well-being and suffering. Sentientist justice intrinsically values sentient life, regardless of its contributions to human society. However, while this approach broadens the scope of justice, it still centers on individual beings, potentially missing the bigger picture of collective ecosystems that sustain those individuals. Ecological justice, in contrast, calls for recognizing entire species and ecosystems as subjects of justice, emphasizing the interdependence of all life forms and the importance of ecological health for the well-being of all. Rethinking the city within this perspective involves more than just granting other individual beings the "right to the city." It challenges us to reimagine our approach to urban design and living. This undertaking is far from straightforward.

A Coyote in My Backyard

In an improbable yet uniquely timed coincidence, as I was writing about the challenges of multispecies equity in urban design, I found myself face-to-face with the very subject of my work: a coyote in my backyard. This unexpected encounter brought the abstract idea to life. Here was Kai (the name my son and I gave the coyote)—a resilient, adaptable creature, navigating the fragmented spaces of a human-dominated landscape. His presence offered a firsthand glimpse into the complexities of coexistence between urban wildlife and human residents, underscoring the urgent need to rethink urban design to make space for all species.

As our eyes met in the clear daylight, I recognized a creature extraordinarily adapted to city life. Kai was standing in a wooded area in my backyard, studying my behavior. Later, I realized it was not merely passing through—it must have found refuge there. In urban areas like Seattle, yards like mine—untamed and unmanaged—act as pockets of biodiversity, small refuges within dense neighborhoods. Coyotes and other adaptable species have started to call these spaces home.

Coyotes, some of nature's most resourceful animals, are increasingly finding their way into suburban and urban areas, pushed there by habitat loss, urban sprawl, and the absence of natural predators. To survive among us, they've changed their breeding patterns, diet, and social behaviors, adjusting to the landscapes we've created. Recent studies reveal genetic differentiation between urban and nonurban coyote populations, suggesting that urbanization shapes evolutionary divergence. However, this differentiation appears to be driven primarily by *nonadaptive* processes, such as genetic drift or isolation, while adaptive evolution has not yet been directly documented.

Urban coyotes exhibit phenotypic adaptations to urban environments, expressed primarily through behavioral and ecological shifts that support survival in human-dominated landscapes. These include increased boldness, nocturnality, and exploratory behaviors, which reshape interactions with humans and enable coyotes to exploit urban resources.[6] Ecological adjustments, such as smaller home ranges and greater reliance on anthropogenic food sources, highlight their capacity for plasticity in resource use within fragmented habitats.[7] Morphological trait changes, such as smaller body sizes, may further facilitate navigation in dense urban landscapes. While the genetic underpinnings of these changes are less well documented, emerging evidence, such as potential resistance to pollutants and diseases in New York City coyotes, points to the possibility of adaptive genetic evolution.[8] These findings underscore the complex interplay of behavioral, ecological, and genetic processes in shaping urban wildlife populations.

However, the coexistence of humans and wildlife often stirs conflict. Wildlife sightings can provoke fear, especially when animals like coyotes are seen as threats to the safety of humans, pets, or properties. Animals, such as raccoons and coyotes, adept at scavenging, have learned to navigate cityscapes, bringing them into closer contact with human communities. Such encounters can lead to tension; humans view wildlife as intruders, disrupting routines or sparking fears of aggression. Without a framework for coexistence, these conflicts tend to escalate, with wildlife cast as pests rather than inhabitants of the urban environment. This dynamic sets up a reactive relationship, in which interactions are often framed around control and removal rather than shared space.

One of my neighbors, for instance, views the coyote as a threat, a reminder that urban design caters to human needs alone. She has even suggested clearing my yard's natural cover to deter wildlife, seeing them as intruders in a human space. This reaction highlights a persistent barrier to multispecies equity: the assumption that cities are only for people. It's a mindset that sees nonhuman species as trespassers when, in reality, we have displaced them. Yet, there is a contrasting trend—a growing curiosity among people about wildlife like Kai. This interest is a valuable asset in the movement to build "multispecies cities"—spaces that embrace our shared ecosystems and the possibility of coexistence.

Varying attitudes toward urban wildlife can lead to disagreements and conflicts among people regarding shared spaces. However, these disputes sometimes create opportunities for mutual learning, paving the way for multispecies coexistence. What began as a disagreement with my neighbor over managing and cohabiting with a coyote evolved into a new relationship forged by our shared experience with wildlife. To foster understanding, I prepared and distributed a flyer explaining why coyotes live in cities and offering guidance on how to safely coexist with them. This simple act sparked curiosity and opened the door to dialogue. The coyote's presence became a catalyst for a new conversation about nature in urban environments. Younger people in particular became fascinated, discovering how interactions between humans and wildlife could positively affect all living creatures in our surroundings.

While curiosity about urban animals is growing, public safety information often stirs fear rather than understanding. Safety messages typically assume people fear wildlife, unintentionally fueling anxieties instead of fostering empathy. Conflicts are framed not as a matter of two species adapting to shared space but as threats—wildlife cast as adversaries in an "us versus them" narrative. This approach overlooks a larger truth: We are the ones who transformed their habitats, removed their predators, and reshaped the land for our benefit. Urban wildlife like Kai are simply navigating a world they didn't design, adapting to survive in landscapes that were once their own. Without framing wildlife safety messages in this context, we risk widening the divide, making coexistence more difficult, and missing an opportunity to nurture empathy for our nonhuman neighbors.

To make multispecies equity a reality, we need more than policy changes; we need a cultural shift. Inclusive cities require reimagining urban spaces as shared habitats, where nonhuman species aren't managed away but respected as part of the community. Kai's presence in my backyard illustrates both the challenge and the opportunity of human-wildlife coexistence in cities. By tapping into people's growing interest in urban wildlife, we may find a path toward cities that acknowledge the right of all species to exist and thrive.

The question is, will we continue pushing wildlife to the edges, or will we adopt a more inclusive vision? Multispecies equity will remain an ideal until we collectively choose to reshape urban life—not just for us, but for all species that help sustain the ecosystems we share.

Rethinking Justice in a Multispecies World

Imagine if justice, as we've traditionally understood it, stretched beyond human concerns. What if our ethical compass expanded to include animals, ecosystems, and entire networks of life, forcing us to rethink who "deserves" justice and why? This idea of multispecies justice is not merely an abstract notion; it's a transformative framework that shakes the foundations of our current theories of fairness, equality, and rights. It suggests that justice need not be a strictly human enterprise but a broader quest to balance the needs of many forms of life.[9]

In rethinking justice for a multispecies world, we aren't just adding a few more voices to the conversation. We're reshaping the very framework of justice itself, moving away from human-centric ideals toward an ethic that respects the interconnectedness of all life. This shift demands not only changes in policy and law but also a profound reimagining of how we view our place in the world.

Historically, justice under modernity strove for universal principles meant to apply equally to all humans, rejecting the discrimination that had plagued society for centuries. Yet, over time, feminist, antiracist, and anticolonial voices revealed cracks in this universalist model, pointing out that it often ignored the fundamental dimensions of race, gender, and other identity-based inequalities.[10] What seemed "neutral" and "fair" often disguised deep-seated biases, favoring a narrow slice of human experiences while sidelining others.

Now, as justice theory increasingly embraces diversity, we find ourselves at a crossroads: Could this embrace of diversity extend beyond human borders? Introducing nonhuman beings as subjects of justice brings its own set of challenges. Animals, plants, and ecosystems possess unique ways of experiencing the world—ways that don't fit neatly within human-centered concepts of fairness and rights. Yet, considering their needs alongside human concerns could lead to a model of justice that's truly inclusive.[11]

Efforts to include nonhuman species in justice frameworks often focus on how their interests can be "represented." But as philosopher Danielle Celermajer argues, this framing is problematic. Representation, especially in its political sense, assumes that someone must speak on behalf of others—typically through human institutions, in human language, using human categories. It presumes that nonhuman beings cannot represent themselves. Yet, as Celermajer points out, many already do. When babies are taken, when rivers stop flowing, when oceans heat up—these are forms of communication about the conditions that interfere with their capacity to function and flourish. "The problem," she explains, "is not that they're not representing themselves—it's that human beings are willfully incapable of attending to those messages." The failure lies not in expression but in perception.

Rethinking representation as inclusion challenges us to reconsider the terms on which beings become visible as political subjects. It also calls for a shift in institutional metaphors: from representation to diplomacy, from speaking *for* to acting *with*. This requires that those standing in for nonhumans—whether scientists, Indigenous knowledge holders, or artists—do so not by translating nonhuman interests into human terms, but by aligning their actions as closely as possible with the ecological realities and relationships of those they aim to support.

Adapting Institutions for a More-Than-Human World

Perhaps the greatest challenge for achieving multispecies justice is adapting our human-centric institutions to accommodate nonhuman needs. Current justice systems, including transitional justice—which often addresses historical wrongs against marginalized human groups—are typically not designed to consider ecological harm or the suffering of nonhuman entities. This absence points to a crucial gap in our institutions.

Celermajer observes, "There is no political system—other than Indigenous political systems—where nonhuman beings participate, or where their interests are included in decisions."[12] Yet, as environmental degradation accelerates, this oversight becomes harder to ignore. Proposals to include nonhuman concerns in these frameworks suggest a radical rethink of what justice means in a world where human and nonhuman lives are intertwined.

One promising concept is "sympathetic imagination." Drawing from philosopher Martha Nussbaum's capabilities approach, this idea encourages us to imagine the experiences of other beings, even those vastly different from ourselves.[13] By fostering an empathetic understanding of how nonhuman life forms interact within their ecosystems, sympathetic imagination could help bridge the empathy gap that humans have toward other species. This approach doesn't aim to humanize other beings but to respect their intrinsic value and their roles within their ecological systems. It suggests a model of justice that honors each organism's contribution to the greater ecological balance.

Extending the concept of justice to nonhumans raises tough questions: How do we represent the interests of nonhuman beings in human-led political systems? Traditional institutions rely on human language, logic, and reason—tools that don't easily translate across species lines. So, while we might strive to "speak" for other species, there's always a risk of misinterpreting or oversimplifying their needs.

These questions are at the heart of my conversation with Celermajer, a professor of sociology and criminology at the University of Sydney, whose work has shaped contemporary debates on multispecies justice. In our interview, she emphasized that multispecies justice requires "moving beyond individual attributes like reason or even freedom understood as non-constrained" and instead focusing on "a range of relational attributes required to support justice—like trust, respect, reciprocity."

Individual Versus Collective Justice

Including nonhuman entities in our justice frameworks isn't as simple as assigning them rights similar to humans.[14] Traditional views of justice are heavily individualistic, focused on the notion of each person—or, in this case, each being—standing on their own. But animals and ecosystems don't operate in isolation; they are deeply interdependent, embedded within webs of ecological relationships. A wolf in a forest, for instance, is not just an individual animal with a set of rights. It's part of a larger system that includes prey, plants, rivers, and countless other organisms. This intricate web of ecological relationships suggests a necessary shift from an individual-rights approach to an ecocentric model. In this perspective, justice extends beyond the needs of each individual species—animal or plant—and instead values the collective health of entire ecosystems. This approach sees ecosystems as intrinsically valuable entities, prioritizing the well-being of the whole over the rights of any single member.

Within the multispecies justice movement, there is an ongoing debate between individual animal rights and the larger ecocentric perspective. On one side, animal rights advocates emphasize the well-being of individual sentient beings. From this view, each animal's capacity to feel pain and pleasure makes it deserving of protection from harm. On the other side, ecological justice advocates argue that prioritizing individuals can sometimes conflict with broader ecosystem health.

As we consider how to move from ethical principles to practices of inclusion, it becomes clear that the challenge is not only structural but also existential. Celermajer refers to the current era as the *Eremocene*—the age of loneliness. "The flip side of human exceptionalism," she explains, "is that we've convinced ourselves we're the only ones who have to come up with the answers, the only ones making sense of this—and that leaves us very lonely." Her reflection names the deeper rupture: a severing of relationship that underlies ecological degradation and social alienation alike. Rejoining a wider community of life requires not only institutional redesign but a transformation in how we understand belonging, attention, and knowledge.

Redefining Equity in Urban Spaces

The idea of multispecies *justice* challenges us to rethink the idea of *equity* in our design approach. What if cities were designed with the needs and perspectives of all species in mind? Multispecies equity requires not only empathy but also a radical reimagining of how we realize multispecies justice in practice. Justice generally refers to the overarching principles and moral foundations that guide how we determine what is "right" or "fair" across all beings, human and nonhuman alike. In multispecies contexts, justice often aims at recognizing the rights or intrinsic values of different species and ecosystems, potentially extending principles like fairness, entitlement, and moral considerability beyond humans to include other forms of life and their environments.

Equity, on the other hand, deals with creating conditions that allow for fairness by acknowledging the specific needs, capacities, and vulnerabilities of diverse entities. In a human context, equity means addressing the specific disadvantages or unique needs of particular groups, ensuring that resources and opportunities are distributed in a way that accounts for these differences. In a multispecies context, equity means recognizing and addressing the differing needs of species and ecological systems to create conditions that support balanced outcomes across interdependent ecological systems.

Around the world, legal systems are beginning to acknowledge the rights of rivers, forests, and ecosystems. Ecuador, Bolivia, and New Zealand have laws granting nature rights. Ecuador led the way in 2008 by embedding the "Rights of Nature" within its national constitution, recognizing the intrinsic rights of natural ecosystems to exist and thrive. Bolivia also passed the "Law of the Rights of Mother Earth" in 2010, granting legal rights to ecosystems, further advancing a holistic legal framework that advocates for the protection of nature as an entity with its own rights. As the Rights of Nature movement gains momentum, cities and local governments are beginning to confront the implications of recognizing ecosystems as entities with inherent rights.

Cities are home to a diversity of species and actively shape ecological and evolutionary interactions that in turn influence urban future. Yet cities are still designed without accounting for other species as legitimate urban inhabitants with distinct needs and vulnerabilities. Planning and design paradigms are largely shaped by a deep-rooted, ontological separation between humans and nature.[15] What would it entail to design cities with the needs and perspectives of other species in mind? Multispecies equity requires not only empathy for nonhuman life but also a fundamental rethinking of how we define justice, governance, and autonomy in an interconnected urban world.

As climate change intensifies and biodiversity declines, the pressure will grow to establish more equitable frameworks that integrate both human and nonhuman interests. Cities will increasingly need to rethink urban planning, governance, and environmental policy to align with principles that recognize the intrinsic value of nature. For local governments, this implies reimagining governance in ways that treat natural systems as stakeholders in decision-making processes, ensuring that development respects ecological integrity and rights of the landscapes and species they impact.

Early signals of this shift are already visible. A recent movement in Washington State to recognize rights for endangered orcas illustrates how multispecies justice could begin to reshape legal and regulatory frameworks. The current US legal system remains heavily anthropocentric, and legal precedents for enforcing the rights of nature are still rare. A greater challenge lies in moving from legal recognition to actionable change in urban planning and design—reframing cities as multispecies ecosystems rather than exclusively human spaces.

This challenge is evident in Skagit County, Washington, where the Sauk-Suiattle Indian Tribe has fought for decades against the impacts of hydroelectric dams on the Skagit River. Built in the early to mid-twentieth century, the dams block salmon migration routes and disrupt nutrient flows essential to the species' survival, violating a US treaty that guarantees the tribe's right to fish. In a landmark "rights of salmon" lawsuit filed in tribal court, the tribe argued that salmon possess inherent rights to "exist, flourish, regenerate, and evolve." The case challenged the City of Seattle's "green" power claims and called for the well-being of salmon and river systems to receive equal consideration. The recent settlement—requiring fish passage around the dams—signals how multispecies equity can begin to reshape infrastructure decisions.

Urban environments are complex ecosystems in which biological and social diversity underpin resilience and adaptability. Yet urban planning has historically prioritized human needs, sidelining other species. As climate change and biodiversity loss accelerate, this narrow framing proves inadequate. A bio-cooperative perspective calls for a redefinition of urban diversity—one that recognizes cities as dynamic systems shaped by enduring interactions between human and nonhuman life, and designs them accordingly.

Operational Autonomy and Interdependence

One of the central principles of a multispecies urban future is operational autonomy. This concept speaks to the design of self-regulating, adaptive systems that account for the diverse needs of multiple species within a shared habitat. In this vision, urban ecosystems must support more than just human life; they must enable all species to function autonomously and collaboratively within a complex web of interactions. Achieving this requires moving away from static, human-centric infrastructure toward systems that can evolve and respond to the needs of different species in real time.

Operational autonomy introduces a new layer of complexity into urban governance, as it demands that the city not only accommodate multiple species but also to sustain the conditions under which they can exercise agency and maintain their own life processes. For instance, urban green spaces can no longer be designed solely for aesthetic or recreational purposes; they must also serve as functioning ecosystems in which other species can persist, adapt, and reproduce on their own terms. This shift raises ethical, practical, and theoretical questions about how governance can recognize species autonomy without ignoring the interdependencies that structure urban life. What ethical frameworks can guide decisions about whose needs take precedence when conflicts arise? And how do we ensure that urban planning does not compromise the autonomy of nonhuman entities in favor of human interests?

Operational autonomy refers to the capacity of a natural environment to regulate itself and maintain its processes without human intervention. This concept has its roots in ecological science, which emphasizes the self-regulating abilities of ecosystems. Ecosystems are capable of cycling nutrients, regulating climate, and supporting a diverse array of species through complex interactions and feedback loops. The paradox of environmental autonomy in multispecies design arises when considering the deep entanglement and interdependence of human and nonhuman species. While multispecies design acknowledges the interconnectedness of life forms, environmental autonomy suggests a degree of separation or independence from human influence. The challenge lies in designing interventions that respect and enhance the autonomy of natural systems while also recognizing the inextricable links between human actions and ecological health. Reconciling environmental autonomy with entanglement and interdependence requires a nuanced approach. Designers can focus on creating conditions that support ecosystem resilience and self-regulation, such as restoring native vegetation, reintroducing key species, and providing space to other beings for self-determination. By doing so, they acknowledge the agency of nonhuman life forms and the inherent capacity of ecosystems to sustain themselves while still considering the reciprocal impacts and dependencies that connect humans and other species. This reorientation also demands what Celermajer calls "attention and humility—not imposing our categories of what it means to relate or to decide—[but being] capable of learning from them."

Toward Multispecies Justice in Urban Design

Expanding the concept of justice beyond the human sphere introduces a new paradigm in urban design and planning. Multispecies justice insists that the well-being of nonhuman species is not an afterthought but an integral part of a city's ethical foundation. In a bio-cooperative city, justice entails not only ensuring that human communities have access to clean air, water, and green spaces but also that nonhuman species are afforded the same considerations. As shared habitats, urban spaces must prioritize the survival and flourishing of all their inhabitants.

A critical component of this approach is the integration of Indigenous knowledge systems. Many Indigenous cultures view nonhuman life forms as active participants in the social and cultural fabric, blurring the lines between human and nonhuman agency. These societies offer valuable insights into how urban planning can adopt multispecies ethics that foster reciprocity and respect. In many Indigenous traditions, the well-being of humans is seen as inseparable from the health of the ecosystems they inhabit. By drawing on these perspectives, urban planners can develop more inclusive models of biodiversity that go beyond simply counting species and instead fostering deep, relational ties between human and nonhuman life.

Two core principles emerge from this multispecies perspective. First, cities must ensure equitable multispecies considerations, embedding the needs of various species into every level of planning. This might include rethinking land-use policies, developing habitats within the urban landscape, or designing infrastructure that supports species migration and adaptation. Equitable urban design must be holistic, recognizing that the survival of one species depends on the health of others and the well-being of the overall ecosystem that maintains multispecies niches.

Second, participatory processes must incorporate nonhuman perspectives in urban planning. While this poses great challenges, novel methodologies are emerging that allow us to understand how nonhuman species interact with urban spaces. It is essential, however, to avoid assuming that humans can fully represent or speak for other species. Such approaches can only approximate nonhuman experiences and must remain grounded in ecological evidence.

Role-play simulations provide one pathway to engage with nonhuman perspectives in planning processes. By crafting scenarios that invite participants to "step into the shoes" of other species, these simulations can deepen understanding of ecological interdependence. When designed with ecological accuracy, they move beyond empathy-building to inform planning decisions by reveling how changes in infrastructure, land use, or management affect nonhuman species. Grounded in empirical ecological knowledge, role-play simulations can expose the constraints and vulnerabilities nonhuman species face in navigating human-dominated environments, offering a structured—though necessarily limited—means of including nonhuman perspectives into the planning process.

As we move toward more inclusive urban planning, participatory processes will need to evolve to reflect a multispecies framework. While humans remain the primary actors in decision-making processes, emerging technologies and practices can help extend human perception to capture the sensory capacities of other species (e.g., light spectra, chemical cues, acoustic frequencies, or microclimatic variation) that inform how nonhuman species experience urban environments. One promising area of development is the creation of multispecies sensors—devices designed to gather real-time urban environmental data from the perspectives of nonhuman entities. For example, sensors that track air quality, humidity, or temperature in the habitats and movement corridors of specific species can offer valuable insights into how urban spaces are experienced by different forms of life.

By embracing the principles of operational autonomy, multispecies justice, and participatory design, urban planners can create spaces that respect the agency of all species. This approach challenges the traditional boundaries of human-centered urban design and offers a vision of cities as inclusive, adaptive ecosystems that support the well-being of all their inhabitants. As cities face increasing environmental challenges, bio-cooperative design offers a pathway to resilience that is both ethical and practical, fostering a future in which humans and nonhumans thrive together.

9 Evolvability, Adaptability, and Foresight

In the heart of Solaria lies the *Garden of the Futures*, a place where the boundaries between what is real and what is imagined dissolve. The garden is not a museum of artifacts or projected ideals—it is a living laboratory, a dynamic, evolving space where experiments unfold, bridging the present, the past, and the futures of the city. Inspired by Italo Calvino's *Invisible Cities*, Fedora, Solaria's *Garden* reverses its narrative.[1] In Fedora, crystal globes hold imagined visions of what the city could have been "if, for one reason or another, it had not become what we see today."[2] Fedora's visions remained in glass—a monument to paths not taken. In Solaria, the living lab invites the city's inhabitants to actively experiment, adapt, and co-create the futures they wish to inhabit.

The garden is spread across the city, embedded in its parks, canals, rooftops, and streets. Each space is a node in a network of ongoing experiments, where visions of the future are living processes. A shaded corridor, planted with trees genetically adapted to withstand extreme heat, serves as a cooling oasis of today's heatwaves while also testing the resilience of urban forests under rising temperatures. Nearby, a neighborhood designed as a food forest allows researchers and residents to evaluate how human and nonhuman communities can thrive together, sharing resources and space. A series of interconnected wetlands, rewilded from an old industrial basin, doubles as a flood buffer and a habitat for returning migratory birds.

What makes the Garden of the Futures unique is its foundational idea: experimenting with the future by embracing uncertainty and designing for adaptability through foresight. Rather than working from a predefined blueprint, the lab operates as a dynamic testing ground, where possibilities are explored, refined, and continuously reimagined. Its philosophy is not about predicting the future with a high degree of certainty but about preparing for a range of possibilities, allowing the city to evolve alongside shifting environmental and social conditions. In this space, uncertainty is not a barrier but a catalyst for innovation, driving adaptive strategies that integrate ecological, technological, and human systems into a cohesive, resilient whole.

Intelligent sensors are embedded throughout the city, capturing air quality, soil composition, and biodiversity in real time. These data streams are fed into open platforms where anyone can explore scenarios and contribute to the experiments. A resident might see how a change in tree canopy and water management would affect their neighborhood, while a scientist might test how pollinator populations respond to new urban planting strategies. In one corner of the garden, bioluminescent fungi pulse faintly at night, responding to changes in soil moisture and offering early warnings of drought conditions. Along a rewilded canal, sensors that are modeled on the electroreception of fish monitor shifts in water salinity and temperature, anticipating potential stressors on aquatic ecosystems. Nearby, a network of acoustic sensors, inspired by the low-frequency sensitivity of elephants, detects distant vibrations, providing critical early warnings of extreme weather patterns. These experiments bridge the sensory worlds of humans and nonhumans, making the garden a dynamic learning system where data become a tool for foresight.

In one space, a microclimate experiment uses bioluminescent plants to provide natural lighting, reducing energy demands. Sensors embedded in pollinator-friendly gardens track the health of bee populations and their impact on food production in nearby urban farms. Another chamber models adaptive water systems inspired by mangroves, where vegetation filters stormwater and floating pathways adjust to seasonal floods. The lab itself is an evolving system, designed to respond to real-time environmental changes. Drawing from biological principles, it incorporates lessons from species like raccoons, which adapt their foraging strategies in urban landscapes, or migratory birds, whose routes shift with climatic cues. Walking through the city, one cannot always distinguish between the lab and the city itself. Wetlands are an experiment and are part of the city's fabric. The elevated walkway, draped in flowering vines that hum with pollinators, is a test of adaptive green infrastructure and a space where people and nature coexist. This intentional blurring of boundaries reminds residents that the city is not an end product—it is alive, always in flux, continuously evolving.

The garden is a space for both envisioning the future and contemplating the past and present. In one part of the city, an old canal has been transformed into an experiment in aquatic rewilding, with fish species that once thrived now returning. The lab also explores an alternate reality, showing what the canal might have been like if it had never been altered. A nearby wall features a projection overlaying the scene with a visualization of Fedora's crystal globes, inviting visitors to consider how futures are shaped not only by paths taken but also by those left unexplored. This interplay between imagination and reality sparks a broader question: How does the way we envision the future shape the present? And how can the present expand the range of possible futures?

Expanding the Boundaries

Environmental change is outpacing cities' ability to adapt, demanding a fundamental transformation in how we integrate nature and urban dynamics. City planners are navigating the urgent and complex task of addressing an increasingly uncertain future, shaped by the convergence of multiple crises such as climate change, biodiversity loss, and social inequalities. To tackle these challenges, they are experimenting with diverse approaches to future planning, including predictive ensembles that use data-driven simulations to project possible outcomes, scenario modeling that explores alternative pathways, and participatory visioning that engages communities in shaping collective visions for the future. Yet, these efforts are often constrained by the biases embedded within their frameworks.

Existing predictive models frequently simplify complex realities, relying on observations and variability of the past to project future trends. Even when these models account for uncertainty, they often define it narrowly—confined to confidence intervals around current trajectories and dynamics. This limits their ability to anticipate transformative change or emergent phenomena, such as tipping points in ecological systems, rapid technological disruptions, and cascading social crises. While scenarios strive to expand the imaginative component of future planning, they often rely on a limited spectrum of paradigms based on skewed and overly simplistic representations of how the world operates. Dominant archetypes—ranging from biological determinism to technoscientific optimism—reinforce narrow worldviews that fail to capture the complexity and diversity of possible futures.

These frameworks do provide valuable insights; however, they are incomplete, capturing only fragments of the broader possibilities. These visions of the future often reflect the present more than they anticipate what lies ahead. They extend existing trajectories, constrained by our current ways of questioning, understanding, and solving problems. Both optimistic and catastrophic narratives remain entrenched in a dichotomous worldview of humans versus nature, or technology versus ecology, leaving them unable to envision or chart a path toward transformative change.

The challenge lies in expanding the spectrum of imaginary archetypes and envisioning futures by exploring emerging new opportunities and actively constructing and testing new possibilities. Can we imagine a future in which humans and natural systems coevolve through mutualistic symbiosis? What would such a world look like, and how can we create it while navigating incomplete knowledge and increasing uncertainty? Understanding and embracing the coevolution of human and natural systems is essential to building resilient societies and transforming our shared habitats into adaptive, thriving ecosystems that reflect a more balanced and integrated vision of cities in nature.

In Solaria, digital twins—data-rich models that simulate how urban ecosystems behave, adapt, and evolve under different conditions—serve as experimental companions to the Garden of the Futures. Drawing on continuous inputs from sensors embedded in plants, soils, animals, and infrastructure, these twins run multiple futures in parallel, modeling how interventions ripple across ecological and social systems. A shift in rainfall patterns, a heatwave, or the loss of a pollinator species can be tested virtually before decisions are made. But more than prediction tools, these digital twins are planning environments—shared platforms where scientists, planners, and residents test hypotheses, debate trade-offs, and adapt strategies in real time. They allow cities to learn from events as they unfold: Feedback becomes foresight. The twin is not a mirror of the city, but an active participant in its transformation.

Unlike Fedora, where futures were trapped in crystal and became mere memories of what might have been, the Garden of the Futures reminds us that futures are malleable. They are experiments in motion, shaped by the actions, choices, and collaborations of those who live in the city today. It is a space where the past and the possible meet, where imagination is not separate from reality but an active force in reshaping it. Through this interplay, the city becomes a living system, continuously reflecting on its trajectory, adapting to change, and coevolving with the people and species that actively shape their shared futures in response to environmental and social change.

The Garden of the Futures is not an imaginarium—a space for creative speculation—it embodies what cities could actually become: dynamic spaces where imagining, experimenting, and future-building occur simultaneously. It serves as a testing ground for ideas, a place where uncertainty drives innovation, and evolutionary potential underpins the adaptability that shapes resilience. This chapter builds on these concepts, exploring how cities can draw from ecological and evolutionary processes to enhance their capacity to adapt and thrive. By understanding how species respond to change and persist under uncertain futures—based on their evolutionary potential—we can reimagine urban environments as dynamic, evolving, and resilient ecosystems in a rapidly changing world.

Evolvability as an Emergent Property

To understand how cities can navigate uncertainty and thrive amid rapid environmental change, we must consider a fundamental, underlying quality of natural systems: evolvability.[3] Evolvability is the capacity of a system to generate adaptive variation, enabling it to respond to novel challenges and opportunities. This property, rooted in ecological and evolutionary processes, provides a powerful framework for designing cities as dynamic, evolving ecosystems.

In living systems, evolvability refers to the ability of populations to produce heritable genetic variation upon which selection can act to generate novel traits.[4] It is not a direct response to environmental pressures but a deeper, emergent property shaped by mechanisms such as genetic recombination, mutation, and developmental plasticity.[5] Evolvability operates over long time scales, often involving multiple generations, and provides organisms with the potential for adaptation—the capacity to innovate and diversify in ways that equip the system to handle future uncertainties. For example, bacteria exhibit evolvability through their rapid mutation rates and horizontal gene transfer, which create a pool of genetic diversity. This enables some individuals to survive antibiotic exposure, even when most of the population is wiped out. The genetic capacity to evolve antibiotic resistance ensures the survival of a lineage under novel and extreme selective pressures, highlighting how evolvability underpins long-term adaptability.[6]

The evolution of Darwin's finches on the Galápagos Islands illustrates how diverse beak morphologies emerged in response to varying food resources. Genetic variation in the finches' beak size and shape provided the raw material for selection, while environmental pressures like droughts or food scarcity drove the survival of those finches with advantageous traits, maintaining their evolutionary potential across generations. Evolvability plays a crucial role in evolutionary success. By enabling populations to produce inheritable variations that natural selection can act upon, it creates a pool of potential adaptations to environmental change. This mechanism provides organisms with the basis for long-term adaptability and resilience, as demonstrated by the finches' ability to endure across generations in a dynamic ecosystem.[7]

Adaptability: A Short-Term Survival Mechanism

While evolvability operates over long time scales, adaptability refers to the capacity of organisms or systems to respond to immediate changes, often through nonheritable, flexible mechanisms such as phenotypic plasticity—the ability of an organism to alter its behavior, physiology, or morphology in response to environmental change.[8] For instance, certain species of fish can adjust their thermal tolerance within a single generation to survive sudden temperature fluctuations in their habitats.[9] These adaptive responses mitigate the immediate impacts of environmental change, providing critical short-term survival mechanisms. Adaptability is critical for buffering populations during periods of rapid environmental change, allowing them to persist until longer-term evolutionary processes can produce heritable adaptations.[10]

Evolvability enhances adaptability, shifting it from a reactive to a proactive capacity, enabling systems to not only respond to immediate challenges but also anticipate and prepare for potential future developments. Adaptability provides the short-term capacity to cope with immediate challenges, while evolvability generates the genetic diversity necessary for populations to adapt over the long term. For example, plants in a fluctuating climate may exhibit adaptability through phenotypic plasticity, adjusting their water-use efficiency during droughts. Over generations, evolvability ensures that genetic variants conferring drought resistance can arise and become widespread, leading to lasting adaptations.[11] This dynamic interplay is amplified through eco-evolutionary feedback loops as species evolve traits that reshape their ecosystems. These traits, in turn, alter the selective pressures on the species themselves and on other organisms, creating conditions that influence the course of future evolutionary trajectories.

Today's warming oceans and frequent bleaching events pose severe threats to coral reefs, as stressed corals expel their symbiotic algae, often leading to mass mortality. Some corals, however, survive by associating with heat-tolerant algae, a short-term adaptive mechanism that helps them endure higher temperatures. This flexibility, while critical for immediate survival, does not alter the genetic framework of the population and is insufficient for long-term resilience in the face of escalating climate pressures.[12]

Corals' long-term survival depends on evolvability, the ability of coral populations to generate genetic variation that fosters adaptation. Genetic differences among individuals allow traits like improved heat tolerance to emerge and spread through natural selection. As these traits become widespread, corals adapt to changing thermal environments, reshaping the reef ecosystem in the process. As corals evolve traits that improve their thermal tolerance, they alter the ecological dynamics of the reef. The structure of the reef changes, influencing fish populations, nutrient flows, and competitive interactions among species. These changes create feedback loops that further shape the evolutionary trajectory of both the coral population and the ecosystem as a whole. This evolutionary feedback highlights the interplay between short-term adaptability and long-term evolvability in shaping ecosystem resilience and transformation.[13]

Evolvability and Adaptability in Cities

The distinction between evolvability and adaptability is crucial when considering how cities can successfully adapt to environmental change. Cities face accelerating pressures from climate change, requiring strategies that address both immediate threats and long-term resilience. Adaptability reflects a city's ability to respond quickly to crises as they arise. After

Hurricane Sandy, for instance, New York City introduced adaptive measures such as raising subway entrances, designing parks that could safely flood, and fortifying coastal defenses. These actions mitigate immediate vulnerabilities, ensuring the city can continue functioning under changing conditions. Adaptability doesn't solve every challenge, but it buys time—allowing urban systems to withstand shocks while developing more innovative, sustainable solutions for the future. The short-term solutions, while critical for immediate survival, must be paired with strategies that enhance evolvability—the capacity to innovate and transform urban systems in response to long-term uncertainties.

Evolvability in urban design parallels ecological principles of maintaining genetic diversity and adaptive potential. For example, cities investing in nature-based solutions, such as planting diverse tree species, not only reduce urban heat islands but also ensure that ecosystems can adapt to changing conditions, such as new pests or climate shifts. Singapore's One Million Trees initiative exemplifies this approach, prioritizing a mix of native and climate-resilient species to restore urban biodiversity and buffer against rising temperatures.[14] Similarly, Melbourne's Urban Forest Strategy focuses on diversifying the urban canopy to increase resilience against drought and disease, enhancing the city's capacity to adapt to future challenges.[15] Along the same lines, coastal cities like Rotterdam demonstrate how adaptability and evolvability can be interwoven through hybrid approaches: Dynamic flood barriers address immediate risks, while ecological measures, such as salt-tolerant wetlands, provide evolving solutions that shift with rising sea levels and changing tidal patterns.[16] These strategies reflect the interplay between short-term flexibility and long-term innovation, a critical balance for urban resilience.

By fostering both adaptability and evolvability, cities can move beyond reactive measures to build systems capable of navigating both current and future challenges. Adaptability allows urban systems to buffer against immediate risks, while evolvability embeds the capacity for innovation and transformation into the design of infrastructure and ecosystems. Together, these principles provide a framework for reimagining urban resilience in the face of climate change, ensuring that cities are not only reactive to present conditions but also equipped to evolve in response to emerging uncertainties.

Evolvability in cities emerges from the interplay of social diversity, institutional flexibility, technological experimentation, and feedbacks between ecological and human systems. These dynamics create the conditions for urban systems to generate diverse solutions and variations that can be tested and refined over time in response to changing challenges. By fostering variation and experimentation, they enable the city to navigate change. For example, Copenhagen's approach to flood resilience includes a variety of adaptive measures, from permeable streets and green roofs to underground stormwater reservoirs.[17] This diversity of interventions reflects the city's capacity to explore multiple pathways simultaneously, maintaining the potential to refine and expand successful strategies.

Several cities across the world are experimenting with strategies aimed at not just responding to immediate challenges but also driving systemic change. Amsterdam, for example, is experimenting with systemic changes through its Circular Economy Action Plan, which prioritizes reducing waste, reusing materials, and transitioning to a regenerative urban economy.[18] Pilot projects range from rethinking construction materials to promoting urban farming, all aiming to shift the city's entire economic and ecological systems toward long-term sustainability rather than addressing single issues in isolation. While combating urban heat is a key goal, Paris's urban forest project is also a broader experiment in transforming how the city integrates nature into its built environment.[19] By introducing green corridors and forests at key urban nodes, Paris is testing how large-scale rewilding might reconfigure urban ecosystems, human health, and biodiversity in tandem with climate mitigation.[20]

Freiburg, Germany, experiments with systemic approaches to urban sustainability, integrating renewable energy, car-free zones, green infrastructure, and co-housing models into a cohesive urban strategy. The city's focus extends beyond immediate challenges like energy efficiency or transport congestion to creating a fundamentally different vision of urban living rooted in social and ecological well-being.[21] New York City's experimentation through initiatives such as the Lower Manhattan Climate Resilience Plan integrates strategies like adaptive flood barriers with a focus on equity and social inclusion. This approach tests how physical and social infrastructures can evolve together to create systemic resilience to climate impacts.[22]

Foresight: Anticipating Future Challenges

Evolution does not anticipate the future; it creates it. Evolvability—an organism's capacity to generate adaptive variation—shapes the conditions for future evolutionary change. Yet whether natural selection can act on this capacity remains a subject of debate. Evolvability is typically considered a characteristic that affects future evolutionary potential. In this light, it may appear "anticipatory" as it relates to an organism's adaptability to unforeseen conditions. This seemingly predictive nature of evolvability challenges the principle of causality in evolutionary theory, which holds that natural selection can only act on present effects, not on future potential that it cannot foresee.[23]

However, recent experimental and theoretical research has provided robust evidence supporting the hypothesis that evolvability—the capacity to generate heritable variation—can itself be a target of natural selection. In one study, researchers exposed populations of *Pseudomonas fluorescens*—a versatile bacterium commonly found in soil and water—to alternating environments that forced the bacteria to repeatedly change their observable traits (phenotypes) to survive. Over generations, these populations evolved mechanisms, such as hypermutation at specific genetic loci, which increased their adaptability. This demonstrated that natural selection can favor genetic architectures that enhance evolvability in response to dynamic environments.[24]

A study on the Lyme disease bacterium *Borrelia burgdorferi* revealed how natural selection shapes its ability to evade immune detection by constantly altering its surface proteins. This adaptability relies on hidden genetic sequences, known as "vls cassettes," which recombine to produce new protein variations. Researchers found strong evidence that natural selection favors changes in these cassettes that increase their genetic diversity, enhancing the bacterium's capacity to evade immune defenses. A critical factor in this process is the presence of highly mutable DNA sequences, called tandem repeats, which further drive genetic variation and are conserved across bacterial strains despite sequence differences. Together, these findings demonstrate how natural selection can act on genetic mechanisms that facilitate rapid and repeated adaptation to environmental pressures.[25]

Another line of evidence comes from theoretical modeling of protein evolution. Simulations demonstrated that under conditions of frequent and severe environmental change, populations with mechanisms favoring evolvability—such as increased genetic recombination or mutation rates—were more likely to succeed. These findings highlight how selection for evolvability becomes particularly pronounced in unstable or rapidly changing environments, where adaptability confers a significant advantage.[26] These studies collectively show that evolvability is not simply a by-product of evolutionary processes but can emerge as a directly selected trait in response to environmental pressures.

There is growing recognition that evolvability can evolve in certain contexts, particularly in rapidly changing or unpredictable environments, where mechanisms enhancing variation provide immediate advantages. However, the extent to which this applies universally or acts as a significant driver of long-term evolutionary dynamics remains a subject of active research and debate. While there is compelling evidence supporting the evolution of evolvability in some systems—as shown above—questions persist about how broadly these findings can be generalized. The idea that natural selection can directly favor traits increasing evolutionary potential is exciting but remains contentious.

A central challenge in this debate is the apparent tension with the principle of causality, as evolvability affects future adaptability rather than providing immediate fitness benefits. Critics argue that natural selection cannot act on traits whose advantages are realized only under future conditions. Additionally, evolvability is typically regarded as a population-level characteristic, subject to less direct selection than traits affecting individual fitness. Empirical studies further complicate the picture by making it difficult to distinguish whether evolvability evolves as an adaptive trait or as a by-product of other evolutionary processes, such as selection for robustness or random genetic drift. These complexities highlight why the evolution of evolvability remains an unresolved and thought-provoking question in evolutionary biology.[27]

Evolutionary change is fundamentally shaped by the current environment and its selective pressures, as populations adapt to immediate challenges to maximize fitness. However, under certain conditions, evolutionary mechanisms can indirectly prepare populations for future environmental changes. For example, bet-hedging strategies like phenotypic plasticity allow individuals to adjust their behavior, physiology, or morphology in response to fluctuating conditions, providing flexibility to navigate uncertain futures. Similarly, genetic diversity within populations serves as a reservoir of traits that may become advantageous under different scenarios, enhancing a population's evolutionary potential.

Other mechanisms, such as evolvability and eco-evolutionary feedbacks, further illustrate how evolutionary processes can set the stage for future adaptability. Traits like modular development or genetic redundancy increase the capacity for rapid adaptation to novel conditions. Additionally, evolutionary changes in one species can alter environmental conditions, indirectly shaping future selective pressures. Keystone species, for example, can transform ecosystems in ways that influence the evolutionary trajectories of other species. In human-modified environments, sociocultural niche construction plays a unique role, as humans create and stabilize selective pressures through actions like urban planning and conservation. These dynamics demonstrate that while evolution is not inherently forward-looking, certain traits and mechanisms can enable populations to adapt to both immediate and future challenges, particularly in the context of predictable patterns or human-driven environmental changes.

In living systems, foresight emerges from mechanisms that enable organisms to hedge against unpredictability. Cities can develop foresight through planning and design strategies that prioritize flexibility and optionality. While adaptability enables a city to remain functional amid immediate change, and evolvability ensures that the city has the capacity to learn, innovate, and diversify its responses over time, foresight provides a mechanism for directing these adaptive capacities toward long-term resilience. Singapore, for example, has integrated foresight into its urban planning by designing water infrastructure that can adapt to both droughts and floods. At the same time, the city-state diversified its water sources through rainwater collection, desalination, water reuse, and imported water, ensuring resilience under varying future scenarios.[28] Foresight enables cities to prepare for multiple possibilities rather than relying on single-purpose, short-term solutions.

Evolvability, adaptability, and foresight are distinct yet interdependent properties that enable living systems—and cities modeled after them—to thrive in changing environments. Evolvability provides the foundation by ensuring that systems can generate diversity and innovation over time.

Adaptability allows for immediate responses to maintain stability, while foresight channels these adaptive capacities toward long-term resilience. The concept of evolvability invites us to rethink urban systems as living entities, capable of generating solutions through ongoing, collaborative experimentation. By drawing inspiration from the evolvability, adaptability, and anticipatory strategies inherent in natural systems, we can draw valuable lessons for reimagining urban environments as dynamic, resilient ecosystems capable of learning, innovating, and responding to uncertainty.

Adaptation Through Multispecies Design

Evolvability, adaptability, and foresight are amplified when seeing the city from a multispecies perspective. Foresight, often associated with human cognition, is increasingly recognized as a widespread trait among nonhuman species. Behavioral flexibility—the ability to adjust actions based on anticipated outcomes—has been observed in insects, mammals, and even arachnids. For instance, grasshoppers and dragonflies use internal models of their environments to guide movement and foraging behaviors, effectively "planning ahead" to maximize resource acquisition. Rats (*Rattus norvegicus*), frequently studied for their cognitive abilities, have demonstrated the capacity to engage their hippocampal regions to simulate navigation in virtual spaces. This imaginative capacity allows them to explore hypothetical scenarios and select optimal paths.[29]

Honeybees offer a striking example of behavioral flexibility in the face of disruption. In experiments conducted more than 200 years ago, the Swiss entomologist Francois Huber showed that bees confronted with obstacles during hive construction could adjust their behavior proactively, as if guided by some sort of master plan.[30] Recent studies suggest that this flexibility reflects their capacity to identify the target of their actions and select the most efficient among multiple envisioned strategies. Although direct evidence for arthropods' neural circuitry supporting the imagination of possible futures is still lacking, neurobiologists Clint Perry and Lars Chittka hypothesize that foresight remains the most plausible explanation for arthropods' ability to solve problems not previously encountered in their species' evolutionary history.[31]

The recognition of nonhuman foresight reshapes our understanding of the role of anticipatory capacity in maintaining resilience. Urban systems, informed by the anticipatory capacities of other species, can adapt to changing conditions more effectively. These natural systems inspire bioengineered solutions that bridge ecological intelligence with urban design.

Harnessing Species' Sensory and Anticipatory Capacities

Understanding how other species sense and anticipate change reveals new possibilities for designing adaptive cities (see figure 1.1 in chapter 1). By examining these sensory modalities, scientists and engineers are developing bio-inspired technologies that reshape how cities monitor, predict, and respond to environmental changes.[32] Intelligent sensors embedded in green infrastructure, for example, monitor ecological health and inform adaptive management strategies.[33] Bioengineered systems inspired by honeybees can detect harmful pollutants in real time, providing early warnings of degraded air quality.[34] Similarly, fire-detection systems modeled on the thermoreceptors of fire chaser beetles enhance wildfire preparedness by identifying ignition points before fires spread.[35]

Some sensory modalities provide even broader applications. Animals like bees and birds, which perceive ultraviolet light, inspire multispectral imaging technologies capable of detecting early signs of environmental stress, such as plant dehydration or nutrient imbalances.[36] Polarized light, used by ants and cephalopods for navigation, informs systems that guide autonomous vehicles and drones in low-visibility conditions, such as during floods or wildfires.[37] These technologies not only mitigate immediate risks but also advance urban systems toward adaptability and resilience by integrating the sensory advantages of other species.

Multispecies-Inspired Foresight and Adaptation

Nonhuman sensory systems are not merely tools for observation—they actively shape the infrastructure of future cities. Bioengineered plants with UV-responsive leaves grow under specific environmental conditions, signaling drought risks or pollution.[38] Air quality sensors inspired by honeybee foraging behaviors translate subtle shifts in atmospheric composition into actionable data, enabling real-time interventions.[39] Similarly, the electroreception capabilities of electric fish have inspired underwater sensors that detect sediment levels or pollution, allowing urban waterways to be managed more dynamically.[40]

The magnetic sensitivity of migratory birds offers another compelling case. By aligning urban planning with geomagnetic pathways, cities can design corridors that support species migration while enhancing human mobility.[41] Acoustic systems modeled on elephants' low-frequency detection provide early warnings of seismic activity, ensuring preparedness for natural disasters.[42] These bio-inspired innovations promote a multispecies perspective, making urban systems functional and sustainable for all inhabitants.

Ultimately, the integration of these sensory systems fosters an evolutionary approach to urban resilience. By combining behavioral foresight with bio-inspired technologies, cities are no longer static constructs but dynamic ecosystems that anticipate challenges, adapt to shifting conditions, and sustain life for both human and nonhuman communities. Together, these innovations redefine urban spaces as living systems—complex, resilient, and equipped to thrive in an uncertain future.

10 Cooperation as a Source of Innovation

Imagine urban trees evolving deeper root systems in partnership with mycorrhizal fungi, enhancing nutrient uptake and resilience to drought. Their canopies cool fast-warming cities through shade and transpiration, while their leaves host beneficial microbes that metabolize pollutants, purifying the city's air. Imagine urban trees cultivating mutualistic alliances with birds—much like acacia trees that host ants in tightly coordinated relationships. The birds defend the trees against herbivores and invasive insects, while the trees provide food, nesting sites, and refuge. Humans, too, join this intricate web of cooperation, nurturing and safeguarding the trees while benefiting from their fruits, nuts, and invaluable ecological services. Over time, this multispecies partnership evolves, ensuring the survival of species and creating biodiverse, resilient urban ecosystems able to withstand the rapidly changing climate.

Urban nature holds the potential of evolving new partnerships that can transform cities into thriving multispecies habitats. In temperate regions, honeybees synchronize their foraging with the changing blooming cycles of urban flora, from wildflowers in green corridors to cultivated plants in community gardens. This mutualistic bond deepens over generations, as gardeners intentionally select resilient flowering species to support pollinator populations. In turn, bees adapt their foraging patterns to align with these flowering schedules, while plants synchronize their blooming cycles to ensure pollination. The interplay between human stewardship, pollinator activity, and plant adaptation strengthens urban biodiversity, bolsters agriculture, and creates a delicate yet robust harmony between human and ecological systems.

In coastal cities, a new partnership between humans, waterbirds, and aquatic vegetation revitalizes urban wetlands. Herons and kingfishers regulate fish and insect populations that might otherwise threaten the aquatic plants critical for water filtration and flood mitigation. Humans, by restoring native vegetation and protecting nesting sites, enable this symbiotic relationship to flourish. Together, these efforts stabilize urban infrastructure, mitigate climate-related risks, and foster refuges for diverse species, illustrating how human–wildlife cooperation can foster urban resilience.

Even in sprawling metropolises, unlikely alliances take root—such as the emerging partnership between rats and ravens. Changes in waste management and shifting food availability draw these two species into closer interaction. Once competitors, they begin to cooperate: Ravens use their intelligence to access hard-to-reach food, while rats scavenge and uncover hidden caches. Over time, this interaction selects for traits that favor mutualism, offering both species survival advantages. Their evolving relationship, shaped by intensifying environmental pressures, exemplifies how species cooperation can drive evolutionary change, reshaping ecological dynamics even in the most inhospitable urban environments.

Cooperative species interactions can drive evolutionary innovations, promote diversification, and open new ecological opportunities. They enable species to access otherwise unavailable resources, increase efficiency, and buffer against environmental fluctuations. Such cooperation can enhance population fitness and abundance, reduce extinction risk, and strengthen adaptive potential. In urban environments—where ecological stressors and rapid change prevail—mutualistic relationships play a critical role in sustaining essential functions such as pollination, nutrient cycling, pest control, and soil formation. This chapter explores how cooperation between humans and nonhuman organisms can be harnessed as an eco-evolutionary strategy to build resilient, bio-cooperative cities.

Evolutionary Foundations of Cooperation

Lynn Margulis showed that cooperation is a fundamental force in evolution through her theory of endosymbiosis.[1] From the emergence of eukaryotic cells to the rise of multicellular organisms and eusocial populations, cooperation has catalyzed major evolutionary transitions. While competition has long been the cornerstone of evolutionary theory, it alone cannot account for the complexity and innovation inherent in biological structures, whether cells, organisms, or eusocial populations. Facilitation and cooperation serve as unifying forces, particularly under environmental stress, enabling the rise of new organizational hierarchies and levels of complexity.

A new, expanded synthesis of eco-evolutionary dynamics is necessary to unravel the conditions and mechanisms underlying major evolutionary transitions. What drives these transitions? Although competition remains a central tenet of evolutionary theory, mounting evidence points to the essential role of positive interactions, such as cellular cooperation within colonies. These interactions foster collective resilience and innovation. While natural selection, rooted in competition, elucidates the diversification within biological levels, it is cooperation that fosters the cohesion necessary for new organizational tiers to emerge. This process becomes especially significant under stressful environmental conditions or within hierarchically structured systems, where cooperation offers adaptive advantages.[2]

As humanity becomes more aware of its role in shaping evolutionary processes, the potential for a transformative relationship with other species emerges. This awareness could catalyze a new paradigm of cooperation, in which humans and other organisms evolve traits that promote mutual benefit and interdependence. Such a shift may signal the emergence of a hybrid planet, where cooperative dynamics operate as an additional force of natural selection, reshaping the biosphere toward greater interconnectedness and resilience.

The Emergence of Cooperation

Cooperation within species, characterized by traits that benefit a group even at individual costs, has long fascinated scientists. Group selection theory suggests that cooperative traits can evolve if they enhance the survival and reproduction of a group, despite potential drawbacks for individuals.[3] Biologists such as E. O. Wilson have championed this idea, arguing that altruistic behaviors can arise through group-level benefits. Eusocial insects like ants provide compelling examples: Worker ants forgo reproduction to support their colonies, a sacrifice that increases the survival and reproductive success of the colony as a whole, ultimately spreading cooperative genes across generations.[4]

Despite its appeal, group selection has faced significant criticism and competing explanations, such as kin selection.[5] Kin selection posits that cooperation arises from genetic relatedness, with individuals promoting the success of relatives to indirectly ensure the propagation of shared genes. The debate between these perspectives has spurred considerable discussion. Proponents of group selection argue that traits fostering group welfare can evolve, even if selfish individuals might undermine cooperation. Critics counter that inclusive fitness and kin selection provide a more robust framework for understanding cooperation, as they account for genetic self-interest within social interactions.

Cultural group selection theory extends the concept into the human domain, emphasizing the spread of cooperative norms and behaviors through cultural evolution rather than genetic mechanisms.[6] According to this view, cooperative cultural traits, such as shared moral codes or collective problem-solving, can enhance group success and propagate through societal imitation and learning. However, some researchers reject group selection entirely, contending that individual-level selection fully explains observed cooperation or that group selection is redundant, offering no additional insight beyond kin selection and reciprocity.

These contrasting theories have fueled ongoing debate and research, reflecting the complexity and multifaceted nature of cooperation both within and across species.[7]

Multilevel Selection

Multilevel selection theory acknowledges that evolutionary processes operate at multiple levels, reconciling individual and group selection by mathematically partitioning evolutionary change.[8] Evidence for group selection provides a compelling argument that evolution extends beyond individual fitness to influence broader biological structures. For example, studies of the "wrinkly spreader" strain of *Pseudomonas fluorescens* demonstrate how specific strains evolve to exploit environmental conditions, persisting in populations despite a selective disadvantage within groups.[9] Similarly, metapopulation experiments reveal how spatial constraints on migration can give a competitive edge to prudent strains—which adopt sustainable, less aggressive strategies that preserve their resources and ensure long-term population stability—over predatory ones. These findings illuminate mechanisms that counteract the "tragedy of the commons," where unchecked exploitation of shared resources can lead to their depletion. Together, these studies reinforce the principles of multilevel selection theory and the role of group selection in shaping microbial behavior and ecological dynamics.

Research across a wide range of organisms, from microbes to vertebrates, supports the role of group selection in evolution.[10] Phenotypic variation among groups and rapid responses to selection in experimental microcosms highlight its significance.[11] Quantitative genetics models show how nonadditive variation within groups, which arises from interactions among genes or individuals that collectively influence traits rather than acting independently, can contribute to additive variation among groups,[12] emphasizing the heritability of group-level traits. For instance, experiments with plants demonstrate that selecting entire groups based on traits like leaf area elicits a stronger evolutionary response than selecting individuals. This suggests that evolution is shaped not solely by individual genetic coding but also by complex interactions within groups. These findings collectively provide robust evidence for the existence and evolutionary importance of group selection.

E. O. Wilson's contributions to the debate on group selection have been particularly influential. Initially a proponent of kin selection, Wilson later challenged this framework in a 2010 paper coauthored with Martin Nowak and Corina Tarnita.[13] They argued that kin selection is neither necessary nor sufficient to explain the evolution of eusociality, the highly cooperative social structure observed in species like ants. Instead, they proposed that multilevel selection offers a more comprehensive explanation: Within-group selection often favors selfish behaviors, even at the expense of the group, whereas between-group selection promotes traits that enhance group fitness, even if costly to individuals. Wilson also emphasized the importance of ecological factors and outlined specific preconditions for the evolution of eusociality, sparking new questions and enriching the debate on cooperation and group selection.[14]

The ongoing global urban transition offers a unique and dynamic context for examining the evolution of cooperation. In cities, the interplay of ecological and evolutionary processes—shaped by shifting environmental conditions, species interactions, and temporal variability—creates novel opportunities for cooperative traits to emerge. Building on multilevel selection theory, it is plausible to hypothesize that traits supporting mutualistic relationships between species may also evolve. Like intraspecific cooperation, mutualism involves interactions that confer reciprocal benefits. Well-documented examples, such as pollination relationships between flowering plants and insects, illustrate how mutualistic traits can evolve in response to shared selective pressures. Yet, coordinating cooperation across species with distinct evolutionary trajectories poses complexities that remain poorly understood and call for further empirical investigation.

Multispecies Cooperation

The hypothesis that cooperative traits are shaped by ecological pressures and multilevel selection mechanisms, though not yet universally accepted, has sparked significant research and debate, offering new perspectives on the complexity of social behaviors in both human and nonhuman species. It has also provided a foundation for exploring cooperative phenomena that extend beyond single species. Evolutionary biologist David Sloan Wilson championed the idea that selection can occur at levels above the individual, proposing that multilevel selection between species is an underexplored but promising area of evolutionary biology.[15] Substantial theoretical and empirical work remains to solidify these ideas within the broader framework of evolution.

Multilevel selection theory rests on two core principles: Selfish behavior can outcompete altruism within groups, but groups in which altruistic members predominate can outperform groups dominated by selfish individuals.[16] Within a single group, selfish traits may provide individual advantages; however, on a larger scale, more cooperative groups often show greater success than less cooperative ones. This framework has become instrumental in explaining the evolution of cooperation within species.

Expanding the concept of multilevel selection to interspecies interactions, mutualism—or cooperation—can be seen as an extension of intraspecific cooperative dynamics. Mutualistic traits may evolve when they enhance the survival and reproductive success of both parties involved. This linkage between intraspecific cooperation and interspecies interactions highlights the potential for multilevel selection to drive mutualistic relationships.

The evolution of mutualistic traits introduces layers of complexity that demand careful consideration of ecological contexts and genetic mechanisms. Mutualistic relationships may develop when the reciprocal benefits of cooperation outweigh the costs for both species. However, interspecific cooperation requires overcoming challenges such as communication barriers, differences in life-history traits, and distinct ecological niches. Each species involved is subject to unique evolutionary pressures, making the formation and maintenance of these partnerships nuanced and context-dependent.

Applying multilevel selection to interspecific cooperation opens exciting frontiers in evolutionary biology, providing a logical extension of existing theories and deepening our understanding of the intricate dynamics within multispecies systems. Exploring these relationships requires integrating insights from ecology, evolutionary biology, and genetics. By framing cooperation within this expanded multilevel selection framework and extending it to human–wildlife interactions, we can uncover new pathways for understanding the selective forces shaping our interconnected world.

The Urban Evolution of Human–Wildlife Cooperation

The emergence of cooperation in both human and nonhuman species is deeply intertwined with ecological contexts. Eco-evolutionary dynamics—the reciprocal interactions between ecological changes and evolutionary processes—create a complex, context-dependent landscape where cooperative traits may evolve. In urban environments, the heterogeneity of resources, community interactions, and shifting temporal dynamics shape selective pressures that pose novel challenges to organisms, potentially encouraging the evolution of cooperative behaviors within species and fostering mutually beneficial relationships between different species.

Urban landscapes—characterized by unique ecological niches and fragmented habitats—serve as incubators for novel cooperative strategies. Species inhabiting these environments often adapt to exploit localized resources efficiently, consistent with theoretical predictions. For example, cooperative behaviors observed in some urban bird species appear linked to resource patterns shaped by human activity. Fragmentation and the scarcity of suitable nesting sites—exacerbated by the combined pressures of urbanization and climate change—present significant challenges. Seasonal fluctuations in resources further create conditions where cooperation becomes not just advantageous but essential for survival. Birds, with their cognitive flexibility and behavioral adaptability, offer a valuable lens for examining how urban environments shape cooperative dynamics. Traits such as reduced body size, enhanced dispersal capacity, broad diets, larger clutch sizes, and increased immunocompetence all contribute to their success in cities. Yet the connection between urban persistence and cooperative behavior raises important questions: Do urban birds exhibit greater degrees of eusociality? Does the success of certain species depend on enhanced cooperation? The answers, it seems, vary widely across taxa.

Take crows, for example. In urban settings, crows often engage in group foraging, an adaptation to fragmented and localized food resources shaped by human activity. This cooperative behavior enables them to efficiently locate and exploit food sources. They also work together to build and defend nests, sharing responsibilities that enhance group success. Additionally, crows communicate about food sources and dangers, such as predators, further illustrating how localized resource patterns drive the evolution of cooperation.

However, not all crow behaviors are cooperative. John Marzluff, an expert on American crows (*Corvus brachyrhynchos*), told me that "American crows are selfish, individualistic actors. They fight each other. They'd kill another [that isn't their mate] if they could catch it on their territory." Yet he acknowledged that "they do cooperate in defending areas, finding food, and scolding predators like eagles when it benefits them all. Cooperation must provide individual benefits for those involved."

Corvid sociality varies widely across species and populations. "Pinyon jays, rooks, and jackdaws are much more cooperative birds," Marzluff explained. Eastern crows, for example, live in larger family units with heightened cooperation, likely influenced by constrained habitat availability. In contrast, Western crows often form pairs with occasional helpers. This divergence highlights how cooperative behaviors may help species adapt to environmental challenges like urbanization and climate change. "Eastern crows, with their stronger social cooperation, may be better equipped to navigate these pressures," Marzluff suggested. Exploring these differences could yield valuable insights into the ecological, behavioral, and genetic factors driving avian social evolution.

Urban environments, through the lens of multilevel selection, can be seen as catalysts for cooperative traits. The fragmented resource patterns of cities create selective pressures that favor group behaviors, which ultimately benefit individuals within these groups. Marzluff elaborated: "In complex urban landscapes, an individual bird may devise a novel adaptation. Cooperation within the group can allow this innovation to spread rapidly." The rapid diffusion of successful strategies enhances group adaptability, facilitating survival and reproduction. Urban settings, therefore, serve as dynamic arenas for individual innovation and group-level adaptation, illustrating the intricate interplay between ecological, behavioral, and evolutionary forces.

Reciprocal Human–Wildlife Cooperation

The fascinating story of the greater honeyguide (*Indicator indicator*) was the inspiration for this book. This extraordinary bird leads humans to wild bees' nests, a behavior long known in Africa and recently confirmed to involve reciprocal signaling between species. In collaboration with a community in the Niassa Special Reserve in northern Mozambique, Claire Spottiswoode and her team study this remarkable mutualistic partnership, in which humans seeking wild bees' nests—known as "honey-hunters"—use a specialized call not used in other contexts, to engage the honeyguides' cooperation. The birds respond with their own calls, in a sophisticated two-way communication system, guiding the honey-hunters to bees' nests by flying from tree to tree. After the beeswax and honey are harvested, the birds feed on the remaining wax combs, completing this extraordinary exchange.[17]

This example of human–wildlife cooperation stands out for its direct reciprocity, deliberate recruitment, and the use of culturally specific human signals with honey-hunting traditions. If honeyguides can cooperate with humans across Ethiopia, Tanzania, Mozambique, and other parts of Africa, might other forms of cooperation still remain undocumented—or new partnerships emerge—in the ever-changing landscapes of urbanized environments?

Over the past decade, Spottiswoode, an evolutionary ecologist and the Pola Pasvolsky Chair of Conservation Biology at the FitzPatrick Institute of African Ornithology at the University of Cape Town, has focused on understanding the conditions under which cooperation evolves and persists in nature. A passionate birdwatcher, she began her research by studying coevolutionary "arms races" between brood-parasitic birds—such as cuckoos and honeyguides—and the host species they trick into raising their young. While her work in southern Africa's savannas initially centered on what she calls "the dark side" of bird life, 12 years ago, she pivoted her focus to explore its brighter side.

Spottiswoode and her colleagues caution that this behavior is under threat. "Human-wildlife cooperation faces unique conservation challenges," they write, "as it requires a motivated human and wildlife partner, a suitable environment, and compatible interspecies knowledge—all threatened by ecological and cultural changes."[18]

In an interview with Spottiswoode and her team, I asked whether she'd considered the potential for new forms of cooperation arising from natural selection, environmental changes, or increased interspecies interactions.

"That's a really interesting question," she said. "We know that mutually beneficial relationships between species—including between humans and other animals—can take a wide variety of forms, and shift quickly depending on ecological conditions." She explained that the form of mutualism with honeyguides is remarkable because of its reciprocal, coordinated behavior. "We define human–wildlife cooperation in a fairly precise way, one which involves coordinated behaviors." She said that such coordinated behaviors, which result in mutually beneficial outcomes, rarely exist between humans and other species. She listed dolphins studied extensively by researchers in Brazil and Myanmar, who appear to provide cues to fishers about when and where fish are available. The fishers then deploy their nets and gear, while the dolphins target the fish that evade the nets.

A potential example of cooperative behavior (on which there are few studies) is ravens (*Corvus corax*), who are believed to signal hunters to their prey in exchange for the opportunity to scavenge the killed animal.[19] New Caledonian crows (*Corvus moneduloides*) seem to indicate to the local Kanak people the site of inaccessible beetle larvae.

Spottiswoode calls the greater honeyguide "the Jekyll and Hyde" of birds. It is both a brutal parasite (deceiving another species into raising its young) *and* a cooperator with humans. She first learned about the "good" side of honeyguides from ornithologist Dr. Hussein Isack, who pioneered work on their behavior among the Boran people of Kenya.[20] Isack's groundbreaking observations included describing not only how honeyguides call to people but also how Boran people call to honeyguides, using a specialized call used only in this context. Twenty years later, Spottiswoode heard accounts of similar two-way communication between honeyguides and Yao honey-hunters in northern Mozambique. She was fascinated and determined to carry out an experimental test of whether this represents reciprocal communication between humans and wild animals.

Using recordings of the Yao honey-hunters' call to recruit honeyguides, her team observed whether honeyguides were able to distinguish the *brrrr-hm* from other arbitrary sounds; that is, could the birds associate the call with a specific meaning? Her experiments, which were carried out in Mozambique's Niassa Special Reserve, experimentally confirmed that not only do honeyguides use calls to solicit human partners, but also that humans use a special call to deliberately recruit a bird's assistance to increase the chances of finding a vital source of food.

Spottiswoode believes that the ability of honeyguides to guide humans is likely genetically inherited—that is, it evolved over hundreds of thousands of years—but is refined through learning in environments in which humans cooperate.

There are many ways that humans interact with animals, such as domestication and training. Domestic animals—like dogs, cats, and horses—are often bred and trained to perform tasks for humans, including hunting, herding, protecting, and transporting. Humans have also trained some animals—like dolphins, whales, falcons, and owls—to assist with tasks or hunt on behalf of humans. Some primates—like chimpanzees and bonobos—have even learned and used human language to communicate and cooperate with us. However, these forms of interaction are not true cooperation between humans and wild, free-living animals because they are not *voluntary* and *mutually beneficial* as are the honeyguide–human coordinated behaviors.

The cooperation between honeyguides and humans is unique because it involves a *non*-domesticated species leading humans to a specific location in exchange for a reward. Honeyguides and humans are able to communicate to achieve *mutually beneficial objectives* that increase the fitness—the ability to survive and reproduce—of individuals of both species. This reciprocal signaling is rare and has been observed in only a few other species. The mutual benefit may have played a role in the evolution of this cooperative behavior, but we do not know how such cooperative human–wildlife behavior has emerged and evolved from other forms of mutualism, nor do we know if there are other undocumented active or novel emerging cases. As Dominic Cram, a colleague of Spottiswoode, pointed out to me, "Reconstructing the evolutionary histories of human–wildlife cooperation is challenging because behavioral traits are not preserved in the fossil record." To address this gap, Spottiswoode's team has developed a framework to contextualize these behaviors, exploring their precursors, evolutionary shifts, and implications for species interactions and ecosystems.

Jessica van der Wal, a behavioral ecologist on Spottiswoode's team, highlights the striking variation in honeyguide–human interactions across Africa. She notes that different communities use distinct sounds—shouting, whistling, chopping wood—to recruit honeyguides, while the birds consistently use a distinctive "chattering" call to signal bees' nests. The team is documenting this diversity to paint a picture of the human landscape of signaling that honeyguides experience. In a recent study, the team showed that human honey-hunting signals used to recruit greater honeyguides vary systematically across regions and are culturally transmitted through social learning, with variation structured by human social organization rather than local environmental conditions.[21]

Spottiswoode's team has recently experimentally tested whether learning plays a role in maintaining a geographic mosaic of honeyguide adaptations to such cultural variation among human communities. In a recent article published in *Science*, Spottiswoode and her colleagues present compelling experimental evidence that honeyguides in Mozambique and Tanzania respectively learn the culturally specific calls of their local human partners: Honeyguides responded more frequently to local honey-hunting signals than to foreign honey-hunting signals or arbitrary human sounds.[22] These findings support the hypothesis that the birds learn to adapt their behavior to local human traditions, strengthening interspecies communication. The study highlights this partnership as a potential case of cultural coevolution, one in which reciprocal learning enables mutual benefits. These insights deepen our understanding of how interspecies cooperation evolves, shaped by both cultural and environmental forces.

The honeyguide's story has reshaped my understanding of human–wildlife interactions. If, as Simon Levin suggests in *Fragile Dominion*, evolution is an open-ended process, it holds the potential for many outcomes—challenging the assumption that human evolutionary success requires the competitive exclusion of other species. This remarkable partnership exemplifies the potential for the emergence of a new partnership between humans and nature. Evolution has produced such extraordinary diversity and complexity of life through multiple forms of interaction among species. A sustainable hybrid planet becomes possible when human–nature relationships are reimagined as true partnerships.

Bio-Cooperation as a Plausible Evolutionary Transition

The question of what drives evolutionary transitions toward greater complexity of life has captivated scientists for a long time. Lynn Margulis was one of the earliest proponents of the idea that complex cells (eukaryotes) emerged through symbiotic relationships between simpler prokaryotic organisms.[23] Building on this concept, theoretical evolutionary biologists John Maynard Smith and Eörs Szathmáry formalized the idea of major evolutionary transitions, defining them as shifts in how information is stored and transmitted.[24] By systematically classifying key transitions—such as gene cooperation forming genomes, the emergence of eukaryotic cells, and the development of multicellular organisms—they demonstrated that, in each case, previously independent entities cooperated to create a more complex system.

Despite advancements in understanding evolutionary change, studying transitions on Earth presents challenges due to the "one-sample experiment" nature of life's history. However, there is growing consensus among scientists that evolutionary transitions involve complex mechanisms where higher-level selection overrides lower-level competition. Revisiting this theory, Szathmáry highlights the central role of multilevel selection in driving major evolutionary transitions, emphasizing how cooperation becomes integral to these profound shifts.[25]

Exploring the emergence of cooperation, alongside its intricate relationship with eco-evolutionary dynamics, offers a fresh—if speculative—perspective on evolution. As Szathmáry and Maynard Smith caution, "There is no reason to expect evolutionary lineages to increase in complexity over time, and no empirical evidence that they do so."[26] Yet the existence of complex life raises compelling questions about how independent agents integrate into unified, higher-order structures. The hypothesis that evolution requires multilevel processes for such transitions provides valuable insight into the conditions needed for these transformative moments in life's history.

Understanding the principles governing biological complexity has implications far beyond theoretical biology, extending into practical applications for human society. E. O. Wilson's revisions to the concept of eusociality illustrate the interplay of selection at different biological levels, linking evolutionary processes to human social structures, cooperation, and governance.[27] Similarly, David Sloan Wilson and his colleagues have applied multilevel selection theory to community dynamics, demonstrating its relevance to fields as diverse as engineering, economics, health, and systems science.[28]

In a compelling synthesis of evolutionary biology and social governance, Wilson, Elinor Ostrom, and Michael E. Cox provide a novel lens on community-based resource management. By drawing parallels between natural selection and the management of common-pool resources (CPRs),[29] they build on Ostrom's pioneering work, which identified eight design principles for effective CPR management.[30] These principles, derived from empirical research on diverse groups, demonstrate that, under certain conditions, communities can sustainably manage shared resources and avoid the tragedy of the commons. Their findings suggest that Ostrom's principles may have broader applications across various group interactions and organizational structures, enhancing our understanding of cooperation in diverse contexts.

Extending the theoretical framework of major evolutionary transitions, the hypothesis that urban eco-evolutionary dynamics could catalyze a new phase of human–wildlife cooperation opens a fascinating direction for inquiry. The complex interactions between humans and other species in urban environments may give rise to emergent properties indicative of a significant evolutionary shift. Under environmental pressures, eusocial organisms—including humans—have evolved remarkable capacities for cooperation. Rapid environmental and social changes could intensify group-level selection, favoring cooperative strategies that benefit entire groups over individual competition.

Such a transition could reshape ecological relationships, enabling new forms of mutualism between humans and other species. While still speculative, the hypothesis is grounded in core principles of evolutionary biology and ecological theory. Urban eco-evolutionary dynamics remain an underexplored scientific frontier—one that calls for empirical research to evaluate the transformative potential of cooperation in shaping the future of life on Earth.

11 Designing Bio-Cooperative Cities

Morning light streams through a lattice of greenery that now, in 2150, defines Paris's skyline, casting intricate shadows over the Seine. From the heights of Montmartre, the city reveals a transformed landscape—a living mosaic of green-roofed buildings and vertical forests interwoven with its centuries-old cathedrals, a testament to the city's coexistence with nature. The green façades cool the summer air and provide sanctuary to birds once disappearing from the city. In parks, residents install nesting boxes and cultivate native shrubs to attract and sustain avian populations. Below, the Seine flows freely, its water revitalized with aquatic plants that oxygenate the river and provide shelter for fish. River otters, long absent from the area, now dart playfully among the roots of floating wetlands engineered with submerged structures and vegetation zones that mimic natural habitats, supporting their reintroduction and survival—an indicator of the river's rebirth.

At Gare Saint-Lazare, commuters gather beneath a sprawling canopy of translucent solar panels intertwined with blooming vines. The panels generate power for the station while offering critical foraging sites for pollinators. Urban gardeners cultivate wildflowers that ensure bees and butterflies can thrive in the bustling urban environment. Trains glide silently along tracks lined with bioswales, where passengers and biodiversity coexist in harmony. Across the city, markets hum with activity—not only from human chatter but from the quiet buzz of bees, the melodies of birds, and the croaks of amphibians. Together, they form a delicate web of life, reinforcing the interconnectedness that sustains this reimagined urban environment.

On the other side of the equator, in Cape Town, morning dew sparkles on the fynbos-covered slopes of Table Mountain, quietly settling into the city below. The transformed urban landscape merges effortlessly with its natural surroundings, reflecting a harmonious integration of city and nature. Rooftop gardens planted with native succulents, like spekboom and red aloe, provide insulation for homes while creating essential corridors for insects and pollinators. At the V&A Waterfront, once solely a commercial hub, floating wetlands anchored with indigenous reeds and rushes, coupled with algae bioreactors, now cleanse the harbor's waters and provide refuge for seabirds and aquatic life. Along the coastline, sprawling kelp forests buffer the city's edge against storm surges while acting as nurseries for marine species that sustain local fishing communities.

In the eastern township of Khayelitsha, innovative urban farms thrive, harnessing closed-loop nutrient recycling systems and solar desalination units to supply fresh food and clean water to a community long challenged by water scarcity. Community greenhouses grow vegetables and medicines, preserving Indigenous traditions that guide their cultivation. In partnership with conservation groups, farmers establish micro-wetlands near their fields, creating habitats for amphibians that aid in pest control and support ecological stability. Under solar-paneled awnings, children play in parks shaded by towering baobab trees, their hollow trunks serving as natural reservoirs to store water for the dry season. Across Cape Town, the coexistence of human and nonhuman life enhances the city's resilience to climate change while honoring its biodiversity and cultural heritage.

Approaching the mid-twenty-second century, cityscapes around the globe have transformed, shaped by the dynamic interplay of climate, culture, and the unique ecological potential of each region. Along South America's tropical coasts, mangrove forests blend seamlessly with urban edges, acting as natural storm barriers against hurricanes and rising tides. The mangroves sustain fish populations, bolster local fisheries, and offer renewable resources such as timber and honey.

In the arid cities of North Africa, the Middle East, and the American Southwest, drought-resistant trees, like mesquite, acacia, and argan, create vital shade canopies that mitigate extreme heat and provide communal gathering spaces. Their deep roots stabilize the soil, curbing erosion and halting the advance of deserts into fertile lands.

Southeast Asia's urban centers embrace their monsoon climates with flood-resilient infrastructure. Floating gardens, wetlands, and permeable surfaces allow cities like Bangkok and Jakarta to manage stormwater effectively while fostering biodiversity. Australia's native vegetation, such as eucalyptus and wattles—adapted to bushfire-prone environments—safeguard Melbourne's communities from wildfires and restore habitats, embodying resilience and renewal.

Technology amplifies nature's potential, creating spaces where ecological function and human ingenuity intersect. Each city's evolution is a collaborative effort—among planners, ecologists, residents, and both human and nonhuman participants. These urban transformations transcend immediate environmental challenges, reflecting a shared commitment to harmonize human civilization with the dynamics of the planetary system. They offer a vision of what is possible when cooperation becomes a guiding principle.

Like a scene from science fiction, the vision of bio-cooperative cities dissolves the boundaries between humans and nature, giving rise to a hybrid planet. From the Seine to the Hudson, and from Cape Town's kelp-lined shores to Bangkok's shaded canopies, cities have evolved into interconnected ecosystems where human and nonhuman life coexist. Once dominated by concrete and steel, these spaces now thrive as zones of collaboration, where conflicts are rare and mutual support is the norm. Has Earth, at the dawn of the twenty-second century, embraced multispecies cooperation?

Walking along Paris's revitalized riverbanks or beneath Cape Town's solar-paneled awnings, one might ask: When did the tide turn? Perhaps it was when humanity stopped seeing cities as separate from the ecosystems that sustain them. Or, as Theo Yoon suggested in his *New York Times* reflection on the 100th anniversary of 2050, it was when we redefined ourselves—not as masters but as participants in nature. This vision of transformed cities shows that change is within reach—an emergent reality rooted in practices and innovations already unfolding worldwide.

Though this vision has yet to be fully realized, its seeds are already visible in cities like Singapore, Freiburg (Germany), Medellín (Colombia), Melbourne (Australia), Curitiba (Brazil), and Cape Town (South Africa). These examples provide compelling evidence that bio-cooperative urbanism is not only possible but a reality already taking shape.

Could Earth be approaching a major evolutionary transition—one defined by an unprecedented partnership between humans and other species? In this emerging paradigm, cities serve as catalysts, fostering interspecies interactions that drive both ecological regeneration and social transformation. Through advanced communication systems and the development of interspecies languages, humans engage with other organisms by adapting signals—vocal, chemical, visual, tactile, and technological—to match the distinct sensory and cognitive capacities of each species.

Cooperation among species has been fundamental to Earth's biological history, shaping the biosphere we inhabit today. Recognizing symbiosis as a driving force of evolution reshapes how we understand and design urban ecosystems. This perspective can inspire a new generation of designers and citizens to build bio-cooperative cities informed by the principles of interspecies cooperation that have shaped life's evolution for billions of years. Such cities would be living models of mutualism, where humans and other organisms coexist and thrive. By aligning urban design with nature's cooperative strategies, these cities could forge a future that mirrors the cooperative origins of life on Earth.

A "City in Nature"

An invitation to join Singapore's Urban Redevelopment Authority (URA) International Panel of Experts offered me a unique opportunity to explore a city long celebrated for its fusion of urban innovation and natural beauty. Singapore is the first city to embrace a bold vision: becoming a "City in Nature." The URA, tasked with shaping Singapore's physical plan for the next 50 years, seeks to transform the city-state from a "Garden City" into one fully integrated with nature—a concept woven throughout its urban fabric.

A year later, after a few virtual meetings during the pandemic, I arrived in Singapore. As the first light of dawn illuminated the city, I embarked on a walk with my son, Matteo, eager to experience this harmonious blend of urban ingenuity and ecological design. Our journey began at the iconic Gardens by the Bay, where the Supertree Grove stood tall against the morning sky. These towering, solar-powered structures, ranging from 25 to 50 meters in height, captivated us with their vertical gardens, serving as refuges for biodiversity and environmental engines for the future. The Supertrees provide shade, generate solar power, act as exhaust air towers, and collect rainwater for irrigation—all while housing more than 162,900 plants representing over 200 tropical species.

Nearby, the Flower Dome and Cloud Forest bio-domes epitomize Singapore's vision of a city that "thinks like a planet." These living masterpieces blend nature, technology, and imagination, demonstrating a sustainable and harmonious approach to urban living. The Flower Dome replicates the dry Mediterranean climate, featuring plants from regions like Australia, South America, and South Africa. Stepping inside transported me back to Calabria, my birthplace, with its familiar scents and dry warmth. In contrast, the Cloud Forest mimics the cool, moist conditions of tropical mountain regions, surrounding the breathtaking 42-meter Cloud Mountain, adorned with orchids, mosses, and ferns. Both bio-domes showcase an intricate balance of biodiversity and technology, inspiring reflection on how cities can coexist with nature.

As we walked along the Marina Bay promenade, marveling at the Sands Hotel with its rooftop garden and infinity pool, Matteo snapped photos of the unfamiliar landscape. He paused and asked, "What would a bird or a bee make of this transformation? Do they see a building or a tree? Will they adapt to this novel environment?" I found myself wondering the same thing.

Our exploration continued at Sungei Buloh Wetland Reserve, one of Singapore's few remaining natural wetlands. Nestled in the northwestern part of the city, its mangroves, mudflats, and freshwater ponds are vital habitats for migratory birds, endangered fish, and a host of other species. These wetlands play a critical role as natural climate regulators, filtering water runoff, reducing pollution, and enhancing environmental quality. Walking along its trails and observation decks, I was reminded of the essential role such spaces play in maintaining urban resilience and ecological health.

Singapore's transformation exemplifies the principles of a "City in Nature," where biodiversity and ecological processes are seamlessly integrated into the urban fabric. A notable example is its rewilding initiative, which repurposed a 24-kilometer former railway line into a "green corridor" connecting forests and nature reserves. This project enhances genetic diversity, supports ecological connectivity, and ensures that every household will be within a 10-minute walk of a park by 2030. It is a bold step toward fostering urban symbiosis.

Walking through Gardens by the Bay, Sungei Buloh Wetland Reserve, and Fort Canning Park provided us with a glimpse into the remarkable interweaving of nature within the cityscape. Singapore's numerous parks and reservoirs—and the Singapore Botanic Gardens, a 160-year-old tropical garden that houses a vast collection of plant species, including the National Orchid Garden—exemplify the city's vision of being a "City in Nature" and its unwavering commitment to maintaining biodiversity.

Singapore's journey is not without its challenges. While its centralized governance has enabled ambitious ecological experimentation, critics argue that the top-down approach limits grassroots-driven innovation. In response, recent policy efforts have begun to expand participatory processes, incorporating greater forms of citizen engagement into planning and environmental initiatives. Singapore's example underscores a powerful lesson: A long-term commitment to ecological integration, research, and experimentation can transform urban spaces into thriving ecosystems.

As Matteo and I ended our day at the waterfront promenade, gazing at the skyline, I reflected on Singapore's achievement in harmonizing modernity with nature. This city's vision offers a compelling example of what may become possible when urban design embraces nature in ways that enable multispecies cooperation.

Determined to understand Singapore's efforts to preserve its natural heritage and foster biodiversity, I interviewed Puay Yok Tan, chief science and technology officer for the National Parks Board and director of the Botanical Garden. Tan, a professor at the National University of Singapore, combines scientific expertise, urban governance experience, and practical collaboration to advance urban greening and ecological preservation. His insights illuminated Singapore's proactive approach to balancing urban expansion with biodiversity.

Curious about the evolution of Singapore's environmental strategy, I asked, "How do 'City in a Garden' and 'City in Nature' differ, and what drove this change?"

Tan explained: "Both concepts aim to make Singapore a green city that enhances livability and resilience to climate change. 'City in Nature,' however, focuses on creating a biodiverse city, not just a green one, and emphasizes connectivity for both human movement and gene flow for flora and fauna by expanding greenways and parks." He added, "This vision is guided by four strategies: extending natural capital, intensifying nature in parks, reintegrating nature into urban spaces, and strengthening ecological connectivity—all supported by a community committed to stewardship."

The City in Nature concept prioritizes restoring biodiversity and ecological processes essential to both wildlife and human well-being, emphasizing resilience and adaptation to climate change. Building on a history of urban ecological research, it incorporates long-term socioecological monitoring to guide initiatives. "We must evaluate outcomes and monitor ecological, biogeochemical, and human–nature relationships," Tan added. This comprehensive approach, supported by the Long-Term Socio-Ecological Research (LTSER) initiative, links urban biodiversity, environmental processes, and social studies to develop sustainable urban ecosystems aligned with the United Nation's Sustainable Development Goals.

The City in Nature concept also anchors Singapore's Green Plan, a national strategy to achieve net zero emissions by 2050. Other pillars include Sustainable Living, focusing on resource efficiency and waste reduction; Energy Reset, advancing renewable energy; Green Economy, promoting sustainable industries; and Resilient Future, enhancing climate resilience with green and blue infrastructure.

The emergence of the City in Nature concept reflects both an innovative mindset and the pressures of rapid environmental change. While it demonstrates Singapore's ability to reimagine human–nature relationships, the centralized governance model may constrain the inclusive green transition needed to address complex challenges. Grassroots initiatives, which often generate creative solutions and public engagement, remain underutilized. Still, City in Nature symbolizes the coevolution of society and nature, exemplifying how urban planning can embed ecological processes into the urban fabric.

Singapore's pioneering efforts in biodiversity integration serve as a global inspiration. Projects like Gardens by the Bay showcase the transformative potential of human–nature partnerships, not just as iconic parks but as symbols of experimental urban ecology. Self-assessment tools, like the Singapore Index on Cities' Biodiversity, which launched in 2008, have provided cities worldwide with frameworks to enhance biodiversity conservation, assess ecosystem functions and ecological performance, and guide governance efforts.

Despite its achievements, questions remain about whether Singapore's model can withstand mounting climate challenges or be replicated elsewhere. Each city faces unique contexts and legacies, making direct emulation impractical. Instead, as Perrine Hamel of Nanyang Technological University notes, Singapore's history of urban ecological experimentation offers inspiration for creating resilient, inclusive cities across the globe.

What is clear is Singapore's unwavering commitment to urban ecological innovation. This legacy underscores the possibility of building biocooperative cities that foster sustainable and harmonious relationships between urban landscapes and the natural world.

The Design of Nature

Urban design's view of nature has evolved over the past century. The Garden City Movement, led by Ebenezer Howard, initially framed green spaces as tools for human well-being.[1] Patrick Geddes later expanded this perspective by advocating for planning that integrated urban areas with their surrounding environments.[2]

The mid-twentieth century brought an ecological shift with figures like Rachel Carson and Aldo Leopold, who emphasized the intrinsic value of nature. Ian McHarg's *Design with Nature* became a landmark in embedding ecological processes into urban planning, seeking harmony between development and conservation.[3] Further advancements in the late twentieth century, led by thinkers like Kevin Lynch,[4] brought the human experience of nature in cities into sharper focus. Kevin Lynch emphasized the sensory and cognitive relationships between urban dwellers and their environment, advocating for the inclusion of natural elements as essential components of urban landscapes.

Building on these foundations, the concept of a *biophilic city*, seeking to seamlessly incorporate nature into urban environments,[5] has become a reference point for ecologically sound urban planning and design. Inspired by E. O. Wilson's concept of biophilia, which denotes the intrinsic human affinity for nature,[6] it was translated into design principles by Stephen Kellert,[7] and articulated by Timothy Beatley as a model of city planning.[8] The biophilic cities movement emphasizes everyday access to nature through green spaces, water elements, and biodiversity, supporting human well-being and ecological resilience.

Regenerative design pushes beyond sustainability by fostering coevolutionary relationships between humanity and the natural world. Rooted in John Tillman Lyle's principles, regenerative design actively enhances ecosystem health by aligning human structures with local ecological and cultural contexts. Using feedback mechanisms and adaptive methods, it promotes ecosystems that are not only sustainable but regenerative, making significant contributions to the environment.[9]

Adaptive design, placemaking, and biodiversity inclusive design (BID) stand out in urban design methodologies for ascribing an active role to biodiversity. BID recognizes species as "citizens" of urban spaces, designing habitats that meet their biological needs and ensure persistence within human-dominated environments.[10]

The shift from designing *with* nature to designing *for* nature is exemplified by novel design frameworks like DeMo (Design and Modeling), which integrates biodiversity's composition, structure, and function into urban planning.[11] This approach creates habitats seamlessly embedded in local ecological networks, supporting both human and nonhuman life. Emerging concepts like the *ecolope*—a multispecies habitat blending built environments with natural systems—illustrate how architecture can foster interspecies relationships. Ecolopes use spatial modeling to accommodate species dynamics, building geometry, and environmental conditions, bridging the human–nature divide and creating inclusive ecosystems that support biodiversity in urban settings.[12]

Multispecies design is grounded in a broader rethinking of agency and relationality in social and ecological theory. Bruno Latour's Actor–Network Theory dissolves the divide between nature and society by recognizing humans, nonhumans, and technologies as co-constitutive "actants" within dynamic networks. Donna Haraway's concept of *sympoiesis*—"making-with"—reframes living systems as collectively produced through ongoing interspecies interactions rather than as bounded, self-organizing units. "More-than-human" thinking brought to prominence in Western scholarship by David Abram and Sarah Whatmore, challenges anthropocentric perspectives by emphasizing the interconnected agencies of all actors—living and nonliving—within ecosystems. Echoing Indigenous knowledge systems, this framework prioritizes collective well-being over human interests and reshapes human–nature relationships by viewing the world as a dynamic, interconnected system.[13]

Multispecies design builds on these ideas, recognizing the reciprocal relationships between humans, animals, plants, fungi, and microbes in shaping environments. It has gained traction in biology, conservation science, and anthropology, particularly through multispecies ethnography, which examines how species influence and co-create cultural and ecological systems. More recent work in planning theory extends these insights by examining how urban environments emerge through the entanglement of human and nonhuman processes.[14] This approach redefines urban design as a collective, system shift, moving beyond human-centered outcomes to engage ecological processes directly.

Despite growing acknowledgment of the need to transcend anthropocentrism, the integration of multispecies principles into urban planning remains limited. Implementation often relegates nature to functional roles or isolated "preserved" spaces, perpetuating a divide between human and ecological systems.

The Vision of Bio-Cooperative Cities

Bio-cooperative cities reimagine the relationship between humans and nature, transforming urban areas into evolving habitats shared by multiple species. These cities recognize the agency of nonhuman life and the complex interdependencies among all organisms. Animals, plants, fungi, insects, and microbes actively shape urban environments alongside humans, making cities products of both human ingenuity and natural processes. Urban design shifts from controlling nature to creating spaces that meet the needs of all life, fostering coexistence and cooperation.

The defining feature of a bio-cooperative city is interspecies cooperation emerging from eco-evolutionary feedback. While biophilic cities seek to reconnect people with nature and regenerative design focuses on ecosystem resilience, bio-cooperative cities place symbiotic interactions between human and nonhuman species at the core of urban ecosystems. In these cities, human and other species not only coexist; they become essential to each other's survival on an increasingly urbanized planet. By moving beyond an anthropocentric focus on human well-being and biodiversity as a resource, bio-cooperative cities seek to create equitable urban spaces for all species.

This vision integrates insights from the social sciences, humanities, and eco-evolutionary dynamics, drawing deeply on Indigenous ecological knowledge. Cities are reconceived as shared multispecies niches, where coexistence and coevolution replace human-centered planning. The shift—from designing with nature for people to designing for nature and people—recognizes the intrinsic value of all organisms and expands urban design to encompass the full spectrum of biodiversity.

By fostering symbiotic interactions and adaptive feedback mechanisms, bio-cooperative cities create systems that support mutual benefit between humans and the biosphere. This perspective links global challenges such as climate change and biodiversity loss to local urban dynamics, spanning neighborhood to regional scales, while accounting for the complexity and heterogeneity of urban environments.

Bio-cooperative cities challenge anthropocentric norms by redefining justice to include the agency and rights of nonhuman species. Urban ecosystems become shared habitats where multispecies equity is central, requiring governance that reflects the complexity of these relationships. Informed by Indigenous ecological knowledge and traditional practices, these cities evolve as co-constructed habitats—dynamic, reciprocal, and resilient.

Principles of Bio-Cooperative Design

The five principles of bio-cooperative design are grounded in the emergent properties of bio-cooperative cities explored in the previous chapters, offering actionable strategies for urban design and planning (table 11.1). Rooted in the complex dynamics of multispecies urban ecosystems, these principles emphasize adaptive interactions, feedback loops, and synergies

Table 11.1
Principles of bio-cooperative design

Principle	Description
Integrate Evolving Ecological Niches	Design urban spaces as interconnected habitats that support the coexistence and ecological functions of diverse species.
Harness Eco-Evolutionary Dynamics	Align urban planning with natural adaptive processes to maintain genetic diversity and adaptive potential and ensure long-term resilience.
Prioritize Multispecies Equity	Recognize the agency, ecological roles, and autonomy of nonhuman species, ensuring that biodiversity thrives alongside urban development.
Enhance Evolvability, Adaptability, and Foresight	Create dynamic urban systems that anticipate, respond to, and evolve in response to unforeseen environmental changes.
Foster Cooperative Innovation	Enable mutualistic relationships between human and nonhuman species to advance innovation in urban governance and resource management.

that redefine cities as dynamic, shared habitats. Their aim is to foster urban environments that enhance resilience and evolutionary potential while balancing ecological and societal needs.

Principle 1: Integrate Evolving Ecological Niches

Bio-cooperative cities envision urban spaces as mosaics of interconnected habitats, where diverse species contribute to the functionality and stability of the ecosystem. This principle focuses on designing spaces that enable species to share overlapping niches, promoting coexistence and mutual growth. It emphasizes the creation of urban environments that actively nurture ecological interactions and processes, enabling species functions. Instead of isolating nature in parks or preserves, bio-cooperative design integrates ecological functions seamlessly into the urban fabric.

Creating effective niches involves designing spaces that promote mutualistic interactions among species. Features such as urban corridors with native plants, green roofs, and bio-receptive building materials provide habitats for pollinators, microbes, and other organisms, allowing them to thrive within human-made environments. These innovations support essential ecosystem functions such as pollination, air purification, and soil stabilization while enhancing both ecological health and urban resilience. By weaving these elements into the urban environment, cities encourage the coadaptation of multispecies within urban ecosystems.

Multispecies niche construction becomes a deliberate tool in urban planning. For instance, constructed wetlands serve dual purposes: managing stormwater and providing critical habitats for aquatic species. Similarly, symbiotic partnerships, such as those between urban trees and mycorrhizal fungi, improve nutrient exchange and boost climate resilience, showcasing the potential of mutualistic design.

To ensure their long-term effectiveness, integrated niches must remain adaptable to ecological changes. Wildlife corridors, bioengineered water systems, and multifunctional urban landscapes offer flexibility, allowing species to adjust to shifting environmental conditions. By embedding adaptability into urban planning, cities can create systems that evolve alongside the organisms they support, ensuring sustained ecological stability.

This principle redefines urban design as a collaborative process aligned with ecological dynamics. By fostering biodiversity, enhancing resilience, and promoting coexistence among species, integrated ecological niches enable cities to function as regenerative ecosystems. In embracing this approach, urban areas can support both human and nonhuman life, meeting the challenges of accelerating environmental change.

Principle 2: Harness Eco-Evolutionary Dynamics

Urban environments create distinctive selective pressures that drive evolutionary adaptations in resident species, profoundly shaping ecosystem functions. Harnessing these eco-evolutionary dynamics involves designing urban systems that anticipate, support, and amplify adaptive processes to enhance resilience and ecological stability.

Designing to preserve evolutionary potential involves maintaining genetic diversity and creating interconnected landscapes that promote gene flow and adaptation. Features such as green corridors and habitat networks enable species to migrate, interbreed, and adapt to urban challenges like heat, pollution, and altered hydrology. These connections mitigate genetic bottlenecks and support the evolution of adaptive traits such as drought tolerance or pest resistance. By treating the city as a dynamic ecological matrix, urban design aligns with and reinforces natural adaptive processes.

Eco-evolutionary feedback loops strengthen resilience by integrating adaptive processes into urban systems. For example, vegetation evolving under urban heat island conditions can mitigate temperature extremes while developing increased stress tolerance. Microbial communities in wetlands adapt to degrade emerging pollutants, exemplifying mutual reinforcement of evolutionary processes and ecosystem functions.

Rapid evolutionary changes often bring trade-offs that impact long-term ecological stability. For instance, plants adapting for rapid growth in urban conditions may compromise root stability, risking ecosystem disruption. Adaptive urban management informed by real-time ecological monitoring—such as genomic tracking and environmental sensors—can preemptively address these trade-offs, ensuring that evolutionary outcomes align with resilience goals.

Nature-based solutions provide dynamic platforms for embedding eco-evolutionary principles into urban planning. Examples include designing urban forests to support species under selective pressure while enhancing air quality, carbon sequestration, and biodiversity; creating bioswales and wetlands to host microbial communities that adapt to filter novel contaminants and maintain water quality; and establishing living shorelines, such as mangroves that evolve with sea-level rise to buffer coastal cities from storm surges. These solutions evolve with environmental changes, addressing emerging challenges while sustaining ecosystem function.

Reframing the city as an evolutionary laboratory positions urban design as an ongoing experiment, a collaborative process between human communities and ecological systems. Cities become living systems, where infrastructure and biodiversity coevolve, enhancing resilience and ecological functionality. Integrating eco-evolutionary principles ensures that cities not only adapt to environmental pressures but also thrive as resilient, dynamic ecosystems.

Principle 3: Prioritize Multispecies Equity

Multispecies equity reframes urban planning by acknowledging the rights and needs of nonhuman species, fostering inclusive ecosystems where biodiversity thrives alongside human development. This principle moves beyond anthropocentric approaches, envisioning cities as shared environments in which the well-being of all species is integral to urban resilience and sustainability.

Inclusive urban design begins with understanding the specific needs and ecological roles of diverse species. By incorporating features such as protected migratory corridors, pollinator-friendly native plants, and wildlife-accessible water systems, urban spaces can support coexistence while enhancing critical ecosystem functions such as pollination and pest control. Tailoring urban spaces to multispecies ecological needs strengthens biodiversity and enhances the resilience of urban ecosystems. Urban infrastructure itself can serve both human and nonhuman inhabitants. For instance, wetlands not only mitigate flooding but also provide essential habitats for amphibians, while urban forests offer cooling benefits for neighborhoods and nesting sites for birds. These systems exemplify how ecological integrity and human functionality can coexist, transforming cities into interconnected networks of mutual reliance between species and their environments.

To achieve this vision, urban governance must include nonhuman species as stakeholders in planning and policymaking processes. Tools such as environmental monitoring systems can track the health of key species, enabling adaptive management practices that respond to ecological needs in real time. Inspired by frameworks such as the Rights of Nature, governance models can incorporate ecological perspectives into urban development, ensuring spaces where all forms of life are respected and valued.

A relational approach to design further highlights the interconnectedness of urban species. Planting vegetation that supports pollinators while enhancing public spaces or protecting culturally significant species reflects a commitment to coexistence. Indigenous knowledge systems, which emphasize respect and harmony with all living beings, offer profound insights into how urban design can align with multispecies values.

By prioritizing multispecies equity, cities can transform into adaptive and ethical ecosystems. Integrating this principle into urban design and governance not only enriches ecological and human communities but also establishes cities as regenerative spaces that thrive in the face of environmental challenges.

Principle 4: Enhance Evolvability, Adaptability, and Foresight

Evolvability is the capacity of a system to generate adaptive variations, enabling it to respond effectively to unforeseen challenges. Designing cities with evolvability in mind involves fostering infrastructure and systems that encourage experimentation, diversity, and innovation. For example, Singapore's One Million Trees initiative diversifies urban forests with native and climate-resilient species, preparing ecosystems to withstand pests, diseases, and climate fluctuations. Evolvable cities operate as dynamic systems, fostering cross-species and interdisciplinary collaboration to adapt and transform over time, aligning urban functionality with ecological resilience.

Adaptability, in contrast, focuses on equipping urban systems to respond flexibly to immediate crises. Post–Hurricane Sandy initiatives in New York City, such as flood barriers and multifunctional green spaces, demonstrate how urban areas can mitigate extreme weather impacts while maintaining functionality. Modular infrastructure and responsive urban systems further enhance adaptability by allowing cities to adjust to changing conditions, supporting resilience during transitional periods, and bridging the path toward long-term sustainability.

Foresight builds on adaptability by integrating long-term resilience into urban planning. This approach involves anticipating diverse potential futures and designing systems that evolve in tandem with environmental changes. Rotterdam's adaptive flood barriers, coupled with ecological restoration, ensure infrastructure remains functional as conditions shift. Similarly, Singapore's diversified water systems address both droughts and floods through redundancy and flexibility, reducing dependence on single-use solutions.

Foresight emphasizes optionality and preparedness, equipping cities to navigate uncertainty while maintaining ecological function and social stability.

Implementing evolvability, adaptability, and foresight requires embedding eco-evolutionary feedback into urban design. Paris's rewilding initiatives exemplify how continuous monitoring of species traits and ecosystem functions can inform adaptive management strategies. Green corridors that facilitate species migration and enhance urban connectivity provide a practical example of how urban infrastructure can align with ecological processes, enabling multispecies coevolution and resilience.

Taken together, evolvability, adaptability, and foresight reorient urban design toward maintaining function through change by building systems able to adjust, diversify, and reorganize over time.

Principle 5: Foster Cooperative Innovation

Cooperation has driven some of the most transformative evolutionary innovations in Earth's history. By leveraging symbiotic relationships among species, urban design can amplify critical ecosystem functions such as pollination, nutrient cycling, water purification, and carbon sequestration. This principle draws from the evolutionary understanding that cooperation fosters resilience and adaptability. Urban systems can emulate these dynamics by creating environments that promote mutualistic interactions. For instance, multifunctional spaces like urban wetlands restored with native vegetation or bioswales planted with pollinator-friendly species demonstrate how ecological functionality can intersect with urban needs. These cooperative landscapes sustain biodiversity, strengthen connections between humans and other species, facilitate symbiotic relationships, and support stewardship based on shared responsibility.

Cooperative innovation must extend beyond ecological interactions to encompass governance systems that promote collaborative planning. Inclusive urban design brings together ecologists, designers, local communities, and policymakers to address the needs of all species. By integrating social and natural sciences, local knowledge, and scientific expertise, cities can ensure more equitable and effective solutions. Participatory approaches, such as co-designing green and blue infrastructure with community input, provide equitable access to resources for both human and nonhuman inhabitants, reinforcing collective ownership and responsibility.

Cities can transform resource management through cooperative systems that align with ecological processes. Composting organic waste to support urban agriculture demonstrates how waste streams can become valuable resources, reducing reliance on synthetic fertilizers and closing nutrient loops. Similarly, incorporating aquatic vegetation into wastewater treatment systems enhances water quality while creating habitats for diverse species. These synergies exemplify intentional designs that foster self-sustaining ecosystems, reduce environmental impacts, and strengthen urban resilience.

Scaling multispecies interactions further magnifies urban resilience. Pollinator pathways linking fragmented green spaces enhance biodiversity and support food security, while urban forests that integrate fungal networks improve ecosystem health and adaptability to climate change. Cooperative pest management strategies, such as encouraging natural predators like birds and bats, reduce dependency on harmful chemicals, fostering healthier urban environments. These initiatives highlight how cooperation among species can amplify ecological and functional benefits, positioning cities as adaptive ecosystems.

Bio-cooperative cities reframe urban systems as dynamic, interconnected networks in which cooperation drives both innovation and resilience. Restoring wetlands to support aquatic life, embedding pollinator-friendly corridors into urban networks, and embracing community-led conservation projects exemplify this vision. These cities thrive on multispecies collaboration, weaving biodiversity into their infrastructure and governance. By fostering partnerships across species and sectors, urban systems evolve into adaptive networks capable of addressing complex challenges, ensuring a sustainable and equitable future for all inhabitants.

From Principles to Practice Through Experimentation

Reimagining urban spaces with a multispecies, multifunctional perspective transforms city planning and design, integrating blue and green infrastructure—such as vegetated roofs, bio-active walls, bioswales, constructed wetlands, and designated open spaces—in ways that facilitate interspecies interactions and support symbiotic relationships. Building multispecies cities requires a fundamental shift in our ontological perspective, our conceptualization of nature, and our methodologies and design practices—from object-oriented to relational approaches and from static to dynamic strategies. Relational design emphasizes interconnected systems rather than isolated entities, recognizing the interdependence between humans and other species. It reimagines space as dynamic and fluid, shaped by interactions and processes rather than rigid compartments. The transition from static to dynamic strategies integrates adaptability, resilience, and coevolution, allowing cities to evolve with changing ecological and social conditions.

This reconceptualization extends earlier critiques of modernist planning. Thinkers like Christopher Alexander and Jane Jacobs questioned the reductive logic of hierarchical design and the neglect of social complexity in urban life.[15] Yet their focus remained largely within a human-centered frame—concerned with spatial structure, social interaction, and lived experience. The approach developed here calls for a deeper ontological and methodological shift: one that positions cities as dynamic, multispecies systems embedded in ecological and evolutionary processes.

This approach reorients urban interventions from managing physical objects to fostering multispecies relationships and ecological processes. Methodologies move away from top-down, standardized planning toward participatory, adaptive approaches that align human needs with ecological systems. The success of design is no longer measured solely by human benefits or economic efficiency but also by biodiversity, resilience, and the capacity to sustain interspecies interactions. By adopting these principles, urban design becomes a process of co-creation—an experimental, evolving practice attuned to the complexity of life on a rapidly changing planet. This integrated perspective links local design practices to global challenges, ensuring that cities function as dynamic ecosystems that support life at all scales.

The concept of bio-cooperative design is not pure imagination but draws on well-documented cooperative interactions in ecological knowledge and empirical evidence. By designing experiments, we can explore the conditions under which these relationships emerge and how they can enhance urban functions. For example, urban design can support cooperative relationships among people, plants, pollinators, and soil biota that sustain reproduction, nutrient cycling, cooling, and water regulation while also addressing human needs. These insights can directly inform elements of multispecies design (figure 11.1).

These are a few examples of potential cooperation between humans and wildlife.

Figure 11.1
Elements of bio-cooperative design illustrating multispecies contributions to urban ecological functioning. Source: Author-created image.

Shade provision. In this scenario, certain plant species could be cultivated or evolve to provide optimal shading in urban areas. Humans recognizing this benefit might engage in daily planting practices, thus facilitating the reproduction of these plants. This relationship creates symbiotic interdependence, balancing human comfort and plant propagation.

Flood protection. Mangroves are known to reduce the impact of storm surges and flooding. Strengthening this capacity may involve both natural evolution and human-assisted development. By selecting and promoting mangrove species with superior protective characteristics, cities near coastlines can enhance their resilience to flooding.

Soil remediation. Microorganisms have been employed in soil remediation, playing a crucial role in breaking down pollutants and rejuvenating soil health in various environments. In urban areas, this concept can be further developed into a more advanced form of cooperation, wherein specific strains of microorganisms, such as bacteria or fungi, are utilized to clean soil contaminated with industrial waste, heavy metals, or other toxic substances, while providing microorganisms with a suitable habitat and nourishment in the form of substances that they can metabolize. By carefully selecting and introducing these organisms in targeted urban locations, soil quality can be improved, and the overall urban ecosystem can be enhanced.

Urban pollination. Pollination, fundamental to plant reproduction, is largely facilitated by insects such as bees. Human-managed beekeeping plays a vital role in ensuring successful pollination in agriculture. By extending this concept to urban areas, initiatives can be developed to strategically place bee colonies and other pollinators in city landscapes. Deliberate planting of flowering plants that provide nectar and pollen can create urban pollination pathways. Humans can actively support these pathways through city planning and community engagement, enhancing ecological connections within the city and ensuring a thriving urban flora.

Pest control. Several species of bats are known to consume vast quantities of insects, including pests. Urban areas that encourage bat populations can benefit from natural pest control, thereby reducing the need for chemical interventions. Cities such as New York have employed falconry as a means of controlling pigeon populations. This approach leverages natural predator–prey relationships to manage bird numbers without harmful chemicals.

Litter control. Many birds, such as raptors and crows, scavenge waste and aid in the removal of organic litter. They act as opportunistic feeders that help clean the city environment by consuming discarded food.

Ecological monitoring. Urban bird populations are indicative of environmental health. Monitoring these populations assists in understanding the ecological status of an area, and this information can be used to guide conservation efforts. For example, crows and ravens can be biosensors for the detection of viruses and other pathogens of public health concern.

Wastewater treatment. Artificial wetlands in urban areas can employ natural processes, in which certain fish and plant species cleanse wastewater. This approach uses ecological principles for sustainable water management.

Climate regulation. Trees planted and maintained in urban areas not only sequester carbon but also provide habitats for various wildlife, creating an interconnected system that benefits both climate regulation and biodiversity.

Disease vector reduction. Encouraging populations of species that prey on disease vectors, such as mosquitoes, can be part of a strategy to reduce the risk of diseases, such as West Nile virus and malaria.

Recreational and therapeutic interaction. Designing urban green spaces that support wildlife not only enhances urban biodiversity but also provides recreational and therapeutic benefits to human inhabitants.

Imagining a Multispecies Urban Future

Building bio-cooperative cities requires redefining humanity's role in Earth's evolution, acknowledging humans as active participants in niche construction and ecosystem dynamics. This vision marks a paradigm shift in urban design, transitioning from human-centric to multispecies planning rooted in a relational and dynamic perspective. By prioritizing relation-oriented over object-oriented philosophies and dynamic over static strategies, urban spaces transform into multispecies ecosystems that integrate blue and green infrastructure to foster interspecies interactions and symbiotic relationships.

Design interventions focus on ecological regeneration and social inclusion, prioritizing biodiversity and resilience. Planning methodologies shift from rigid master plans and top-down approaches to adaptive management, participatory design, and feedback loops informed by ecological monitoring. Technology leverages multispecies sensory and anticipatory capacities to create responsive networks for early warning and adaptive strategies in response to climate change. Success is redefined, measured not by economic growth or infrastructure efficiency but by ecological health, interspecies relationships, and sustained biodiversity.

Bio-cooperative urban design rests on four key pillars: fostering interconnectedness among species, creating dynamic and adaptive urban spaces, embracing complex governance structures, and adopting relational-oriented infrastructure. This approach prioritizes multispecies interactions, flexible design strategies that evolve with ecological and social needs, governance systems capable of addressing the complexity of multispecies well-being, and infrastructure that functions as integrated, multifunctional systems. Together, these elements orient urban design toward the conditions that allow human and nonhuman communities to persist, interact, and adapt over time.

To actualize the vision of bio-cooperative cities, it is necessary to acknowledge our integral part in nature's grand plan. The transformative shift in human consciousness outlined here might seem unlikely, even unimaginable, unless considered through the lens of evolution. Emerging insights from population genetics studies of niche construction suggest that niche construction can reshape evolutionary trajectories and stability, leading to novel evolutionary dynamics and coevolution. I propose that our current epoch presents a unique opportunity for systemic change, prompted by the convergence of major transitions in both human society and Earth's systems. This confluence could trigger a radical transformation in our relationship with nature if humanity has the imagination to envision a multispecies urban future.

Conclusion: Multispecies Symbiosis as an Eco-Evolutionary Strategy

As we face the converging crises of climate change, biodiversity loss, and urbanization, the vision of bio-cooperative cities offers a transformative pathway to resilience and sustainability. Grounded in a multispecies perspective and guided by principles of interspecies cooperation, these cities challenge anthropocentric urban planning. They reimagine urban spaces as shared habitats where humans and other organisms co-create the conditions that enable life on Earth to thrive over the long term.

The examples in this book—from Singapore to Cape Town—demonstrate that integrating ecological and evolutionary principles into urban design is not only possible but essential for addressing today's challenges. These pioneering efforts reveal how cities can lead an eco-evolutionary transformation, fostering interspecies interactions and feedback loops that sustain biodiversity, cultural vitality, and planetary health.

This book calls on all who are involved in city building to adopt a new paradigm: one in which human ingenuity aligns with nature's cooperative strategies, transforming urban environments into catalysts for eco-evolutionary resilience. By embracing multispecies design principles, we can reimagine cities as living systems, capable of adapting to environmental change and ensuring a thriving planet for generations to come.

This journey began with a question: How can we expand our vision of the future to imagine a world where humans collaborate with other species? Inspired by the honeyguide's cooperation with honey hunters in Mozambique, I asked: Could rapid urbanization and eco-evolutionary dynamics spark new forms of human–wildlife cooperation? What might such a world look like?

Reflecting on this question, I return to Africa—a continent rich in ecological wisdom and cultural narratives often sidelined in global discourses.[1]

Africanfuturism, deeply rooted in Indigenous knowledge and bold reimagining of the Anthropocene, offers pathways to futures that are both sustainable and deeply connected to the diversity of life. This perspective underscores that the future is multifaceted, shaped by a diverse interplay of culture, environment, and tradition.[2] This viewpoint challenges Western-centric models of urbanization, urging us to embrace the richness of planetary stories. Africanfuturism exemplifies this approach by merging ancient traditions, ecological knowledge, and innovative visions to redefine resilience—not as a static goal but as a dynamic process that celebrates diversity, adapts to change, and fosters symbiosis.[3]

As political ecologist Henrik Ernstson and environmental historian Sverker Sörlin argue in *Grounding Urban Nature*,[4] reimagining urban nature requires resisting the simplification of a singular global model. Instead, we must draw on the rich, localized socioecological narratives, as Africanfuturism and other place-based perspectives highlight, honoring the complex interplay of cultural, social, and environmental factors in urban ecology.

Expanding our imaginative capacity through diverse narratives strengthens our scientific understanding and inspires designs for resilient, adaptive, and sustainable cities. While storytelling traditions from Africa exemplify the power of such narratives, the core message transcends any single culture. It is about embracing inclusive, multifaceted approaches that acknowledge multiple pasts and conceptual frameworks, opening new pathways for innovative urban planning and sustainable living in harmony with nature.

The transformative shift in human consciousness required for this vision may seem daunting, yet evolutionary theory helps clarify how such change might unfold. Research on niche construction and coevolutionary dynamics suggests that systemic innovation and stability can emerge through reciprocal interactions between organisms and environments. This rare convergence of societal and ecological transitions presents an unparalleled opportunity for bold, systemic change.

The path forward demands collective imagination, action, and the humility to see ourselves not as separate from nature but as integral participants in its ongoing story. By aligning urban development with nature's cooperative strategies, we can co-create regenerative cities that sustain life.

The question is no longer whether we can achieve this change but how quickly and effectively we can act. Embedding urban development within nature's cooperative framework unlocks the potential for humans and the planet's diverse life forms to flourish together.

This vision is not just a blueprint for the future; it is an urgent call to action.

The work begins now.

Glossary

Adaptive evolution: Change in the frequency of beneficial genetically based phenotypes in response to natural selection.

Allele frequencies: The proportion of different versions of a gene (alleles) present in a population, which can change over time due to evolutionary pressures.

Anthropocene: A proposed geological epoch that acknowledges the significant impact of human activities on Earth's ecosystems and climate.

Bio-cooperative cities: Urban areas designed with the intention of fostering cooperative relationships between humans and other species to sustain ecosystem functions and promote well-being.

Criticality in dynamic systems: A concept from physics and systems theory describing a system on the brink of a phase transition, believed to optimize certain computational properties and responsiveness.

Cultural coevolution: The reciprocal and dynamic evolutionary process in which human culture and biological traits influence and adapt to each other.

Developmental bias: The tendency for certain kinds of variations during development to occur more frequently than others, influencing the direction of evolutionary changes.

Divergent evolution: The accumulation of differences between closely related populations or species, leading them to become more distinct over time.

Eco-evolutionary coupling: The interconnectedness of ecological interactions and evolutionary changes, where each influences the other over time.

Eco-evolutionary feedbacks: Reciprocal interactions between ecological and evolutionary dynamics on contemporary time scales.

Ecosystem: A unit comprising all of the organisms in a given area interacting with the physical environment, so that a flow of energy leads to a clearly defined trophic structure, biotic diversity, and material cycles (i.e., exchange of materials between living and nonliving parts).

Ecosystem function: The biological, geochemical, and physical processes occurring in an ecosystem that collectively regulate the flux of energy, organic matter, and nutrients.

Ecological niche: The functional role and position of a species in the ecosystem, including what resources it uses and how it interacts with other species.

Ecological processes: The natural procedures and interactions that maintain life and biodiversity in ecosystems, such as pollination, nutrient cycling, and food chain dynamics.

Evolutionary change: Alterations in the genetic makeup of a population over successive generations, often leading to the development of new species.

Evolutionary potential: The capacity of a species or population to undergo evolutionary change in response to selective pressures over time.

Evolutionary trade-offs: Situations in which the evolution or development of one trait may be linked to the loss or reduction of another due to the limitations of resources or genetic correlations.

Evolutionary transition: Significant shifts in the organization and complexity of life, such as the emergence of multicellular organisms, which affect the trajectory of evolution.

Great Oxidation: A major event in Earth's history, occurring approximately 2.4 billion years ago, characterized by a significant increase in atmospheric oxygen. This event, primarily driven by the photosynthetic activity of cyanobacteria, led to profound changes in the planet's environment, paving the way for the evolution of aerobic life forms.

Human ecological niche: The role and space that human beings occupy in the ecosystem, including their use of resources and their interactions with other species.

Interspecific hybrids: Offspring resulting from the mating of two different species, which may display characteristics of both parents or new traits.

Multispecies cooperation: Collaborative interactions between multiple species, which may include mutual support and resource sharing.

Niche construction: The process by which organisms alter their own and each other's environments, often affecting their own and other species' evolution.

Pace-of-life syndromes: Variations in life history strategies, including growth rate, lifespan, and reproduction, that are influenced by ecological conditions.

Phenotypic change: Observable and measurable changes in the physical and biochemical characteristics of organisms, which may be influenced by genetic and environmental factors.

Phenotypic plasticity: The capacity of an organism to alter its physical or behavioral characteristics in response to varying environmental conditions.

Planetary evolution: The changes and developments occurring on a planetary scale, influenced by a combination of natural processes and human activities.

Resilience: The ability of a system to absorb disturbances and reorganize while undergoing change so as to retain its essential functions, structures, and feedbacks.

Robustness and flexibility: In evolutionary biology, robustness refers to an organism's or system's ability to persist despite environmental or genetic perturbations. Flexibility denotes the capacity of an organism or system to adaptively change in response to these perturbations.

Species symbioses: Close and long-term biological interactions between different species, including mutualism, commensalism, and parasitism.

Symbiotic relationships: Interactions between different organisms that live closely together, often benefiting one or both involved parties.

Systemic change: Fundamental changes that affect or pertain to an entire system, rather than isolated parts.

Taxa: Groups of one or more populations of organisms classified by scientists to form a unit, such as species, genus, and family.

Urban ecosystems: Ecosystems comprising both human (social, economic, and built environment) and natural (flora, fauna, and physical) components, characterized by intense human activity and management, high resource input, and significant outputs of waste and pollution.

Urban phenotype: The characteristic traits of organisms that have evolved in response to urban environments.

Urban selection pressures: Specific environmental and social factors in urban areas that influence the survival and reproduction of organisms.

Urban sustainability: The ability of urban areas to maintain ecological and social systems in a healthy and viable state over the long term.

Urban-driven evolutionary change: Evolutionary alterations in species that occur specifically due to pressures and conditions found in urban environments.

Urbanization: The process by which towns and cities are formed, characterized by an increase in population density and the development of physical infrastructure.

Notes

Introduction

1. Jan Zalasiewicz, Colin Waters, and Martin J. Head, "Anthropocene: Its Stratigraphic Basis," *Nature* 541, no. 7637 (January 18, 2017): 289; Colin N. Waters et al., "The Anthropocene Is Functionally and Stratigraphically Distinct from the Holocene," *Science* 351, no. 6269 (January 8, 2016): aad2622.

2. Erle C. Ellis, "Ecology in an Anthropogenic Biosphere," *Ecological Monographs* 85, no. 3 (2015): 287–331.

3. Jonathan W. Atwell et al., "Boldness Behavior and Stress Physiology in a Novel Urban Environment Suggest Rapid Correlated Evolutionary Adaptation," *Behavioral Ecology: Official Journal of the International Society for Behavioral Ecology* 23, no. 5 (September 2012): 960–969; Jenélle L. Dowling, David A. Luther, and Peter P. Marra, "Comparative Effects of Urban Development and Anthropogenic Noise on Bird Songs," *Behavioral Ecology: Official Journal of the International Society for Behavioral Ecology* 23, no. 1 (2011): 201–209; Pierre-Olivier Cheptou et al., "Rapid Evolution of Seed Dispersal in an Urban Environment in the Weed Crepis Sancta," *Proceedings of the National Academy of Sciences of the United States of America* 105, no. 10 (March 11, 2008): 3796–3799; Brooks E. Miner et al., "Linking Genes to Communities and Ecosystems: Daphnia as an Ecogenomic Model," *Proceedings of the Royal Society B: Biological Sciences* 279, no. 1735 (2012): 1873–1882; Kristien I. Brans et al., "The Heat Is On: Genetic Adaptation to Urbanization Mediated by Thermal Tolerance and Body Size," *Global Change Biology* 23, no. 12 (2017): 5218–5227; Kayla D. Coldsnow et al., "Rapid Evolution of Tolerance to Road Salt in Zooplankton," *Environmental Pollution* 30 (2016): 1e7.

4. Kristien I. Brans, Robby Stoks, and Luc De Meester, "Urbanization Drives Genetic Differentiation in Physiology and Structures the Evolution of Pace-of-Life Syndromes in the Water Flea Daphnia Magna," *Proceedings: Biological Sciences* 285, no. 1883 (July 18, 2018): 20180169; Amy K. Hahs et al., "Urbanisation Generates Multiple Trait Syndromes for Terrestrial Animal Taxa Worldwide," *Nature Communications* 14, no. 1 (August 7, 2023): 4751.

5. Eric P. Palkovacs et al., "Fates Beyond Traits: Ecological Consequences of Human-Induced Trait Change," *Evolutionary Applications* 5, no. 2 (2012): 183–191; Marina Alberti, "Eco-Evolutionary Dynamics in an Urbanizing Planet," *Trends in Ecology & Evolution* 30, no. 2 (February 2015): 114–126.

6. G. Evelyn Hutchinson, "Concluding Remarks," *Cold Spring Harbor Symposium on Quantitative Biology* 22 (1957): 415–427.

7. United Nations, "Orders of Magnitude of the World's Urban Population in History," January 1, 1976, https://www.un.org/development/desa/pd/content/orders-magnitude-worlds-urban-population-history.

8. United Nations Publications, "World Urbanization Prospects: The 2018 Revision" (UN, 2019), https://population.un.org/wup/publications/Files/WUP2018-Report.pdf.

9. Marina Alberti, "Cities of the Anthropocene: Urban Sustainability in an Eco-Evolutionary Perspective," *Philosophical Transactions of the Royal Society B: Biological Sciences* 379, no. 1893 (January 1, 2024): 20220264.

10. Alberti, "Cities of the Anthropocene."

11. David Pimentel, "Animal Population Regulation by the Genetic Feedback Mechanism," *The American Naturalist* 95, no. 881 (1961): 65–79; David M. Post and Eric P. Palkovacs, "Eco-Evolutionary Feedbacks in Community and Ecosystem Ecology: Interactions Between the Ecological Theatre and the Evolutionary Play," *Philosophical Transactions of the Royal Society of London: Series B, Biological Sciences* 364, no. 1523 (June 12, 2009): 1629–1640; Thomas W. Schoener, "The Newest Synthesis: Understanding the Interplay of Evolutionary and Ecological Dynamics," *Science* 331, no. 6016 (January 28, 2011): 426–429.

12. Schoener, "The Newest Synthesis."

13. James S. Santangelo, L. Ruth Rivkin, and Marc T. J. Johnson, "The Evolution of City Life," *Proceedings: Biological Sciences* 285, no. 1884 (August 15, 2018): 20181529.

14. Vasilis Dakos et al., "Ecosystem Tipping Points in an Evolving World," *Nature Ecology & Evolution* 3, no. 3 (March 2019): 355–362.

15. Axel Kleidon, "Working at the Limit: A Review of Thermodynamics and Optimality of the Earth System," *Earth System Dynamics* 14, no. 4 (August 30, 2023): 861–896.

16. Adam Frank, Axel Kleidon, and Marina Alberti, "Earth as a Hybrid Planet: The Anthropocene in an Evolutionary Astrobiological Context," *Anthropocene* 19 (September 1, 2017): 13–21; Marina Alberti, *Cities That Think Like Planets: Complexity, Resilience, and Innovation in Hybrid Ecosystems* (Seattle: University of Washington Press, 2016).

17. Timothy M. Lenton and Hywel T. P. Williams, "On the Origin of Planetary-Scale Tipping Points," *Trends in Ecology & Evolution* 28, no. 7 (July 2013): 380–382.

18. Marina Alberti et al., "Global Urban Signatures of Phenotypic Change in Animal and Plant Populations," *Proceedings of the National Academy of Sciences of the United States of America* 114, no. 34 (August 2017): 8951–8956.

19. Marina Alberti et al., "The Complexity of Urban Eco-Evolutionary Dynamics," *Bioscience* 70, no. 9 (September 18, 2020): 772–793.

20. Marina Alberti et al., "Evolving Nature-Based Solutions for Urban Resilience," *Science* 392, no. 6799 (May 14, 2026): 701–708.

21. Lynn Margulis and René Fester, *Symbiosis as a Source of Evolutionary Innovation: Speciation and Morphogenesis* (Cambridge, MA: MIT Press, 1991).

22. Dominic L. Cram et al., "The Ecology and Evolution of Human-Wildlife Cooperation," *People and Nature (Hoboken, N.J.)* 4, no. 4 (August 2022): 841–855.

23. Alberti, *Cities That Think Like Planets.*

24. Kevin Laland et al., "Does Evolutionary Theory Need a Rethink?" *Nature* 514, no. 7521 (October 9, 2014): 161–164.

25. Christian Torres-Sosa, Sui Huang, and Maximino Aldana, "Criticality Is an Emergent Property of Genetic Networks That Exhibit Evolvability," *PLoS Computational Biology* 8, no. 9 (September 6, 2012): e1002669.

Chapter 1

1. Jakob von Uexküll, "A Stroll through the Worlds of Animals and Men: A Picture Book of Invisible Worlds," *Semiotica* 89, no. 4 (1992): https://doi.org/10.1515/semi.1992.89.4.319. For a recent popular treatment, see Ed Yong, *An Immense World* (New York, NY: Random House, 2022).

2. Miner et al., "Linking Genes to Communities and Ecosystems: Daphnia as an Ecogenomic Model"; Brans et al., "The Heat Is On"; Coldsnow et al., "Rapid Evolution of Tolerance to Road Salt in Zooplankton."

3. Cheptou et al., "Rapid Evolution of Seed Dispersal in an Urban Environment in the Weed Crepis Sancta."

4. Alexander V. Badyaev et al., "Evolution on a Local Scale: Developmental, Functional, and Genetic Bases of Divergence in Bill Form and Associated Changes in Song Structure between Adjacent Habitats," *Evolution: International Journal of Organic Evolution* 62, no. 8 (August 2008): 1951–1964.

5. Jane Andre et al., "Molecular Genetic Differentiation in Earthworms Inhabiting a Heterogeneous Pb-Polluted Landscape," *Environmental Pollution* 158, no. 3

(March 2010): 883–890; Craig Anderson et al., "Genetic Variation in Populations of the Earthworm, Lumbricus rubellus, across Contaminated Mine Sites," *BMC Genetics* 18, no. 1 (November 17, 2017): 97.

6. Noah M. Reid et al., "The Genomic Landscape of Rapid Repeated Evolutionary Adaptation to Toxic Pollution in Wild Fish," *Science* 354, no. 6317 (December 9, 2016): 1305–1308.

7. Post and Palkovacs, "Eco-Evolutionary Feedbacks in Community and Ecosystem Ecology: Interactions between the Ecological Theatre and the Evolutionary Play."

8. Stephen R. Carpenter, James F. Kitchell, and James R. Hodgson, "Cascading Trophic Interactions and Lake Productivity," *Bioscience* 35, no. 10 (November 1985): 634–639; Stephen R. Carpenter and James F. Kitchell, "Consumer Control of Lake Productivity," *Bioscience* 38, no. 11 (December 1988): 764–769; Erik Jeppesen et al., "Shallow Lake Restoration by Nutrient Loading Reduction—Some Recent Findings and Challenges Ahead," *Hydrobiologia* 584, no. 1 (June 2007): 239–252.

9. Nelson G. Hairston Jr. et al., "Rapid Evolution Revealed by Dormant Eggs," *Nature* 401, no. 6752 (September 1999): 446–446; Orlando Sarnelle and Alan E. Wilson, "Local Adaptation of Daphnia Pulicaria to Toxic Cyanobacteria," *Limnology and Oceanography* 50, no. 5 (September 2005): 1565–1570; David M. Post and Eric P. Palkovacs, "Eco-Evolutionary Feedbacks in Community and Ecosystem Ecology: Interactions between the Ecological Theatre and the Evolutionary Play," *Philosophical Transactions of the Royal Society of London. Series B, Biological Sciences* 364, no. 1523 (June 12, 2009): 1629–1640.

10. E. van Donk and R. D. Gulati, "Transition of a Lake to Turbid State Six Years after Biomanipulation: Mechanisms and Pathways," *Water Science and Technology: A Journal of the International Association on Water Pollution Research* 32, no. 4 (August 1, 1995): 197–206; Ramesh D. Gulati and Ellen van Donk, "Lakes in the Netherlands, Their Origin, Eutrophication and Restoration: State-of-the-Art Review," in *Ecological Restoration of Aquatic and Semi-Aquatic Ecosystems in the Netherlands (NW Europe)* (Dordrecht: Springer Netherlands, 2002), 73–106; Erik Jeppesen et al., "Biomanipulation as a Restoration Tool to Combat Eutrophication," in *Advances in Ecological Research*, Advances in Ecological Research (Elsevier, 2012), 411–488.

11. Cheptou et al., "Rapid Evolution of Seed Dispersal in an Urban Environment in the Weed Crepis Sancta."

12. James M. Bullock et al., "Human-Mediated Dispersal and the Rewiring of Spatial Networks," *Trends in Ecology & Evolution* 33, no. 12 (December 1, 2018): 958–970.

13. Monika von der Lippe and Ingo Kowarik, "Long-Distance Dispersal of Plants by Vehicles as a Driver of Plant Invasions," *Conservation Biology* 21, no. 4 (2007): 986–996.

14. Andre et al., "Molecular Genetic Differentiation in Earthworms Inhabiting a Heterogeneous Pb-Polluted Landscape."

15. Anderson et al., "Genetic Variation in Populations of the Earthworm, Lumbricus rubellus, across Contaminated Mine Sites."

16. Sarah E. Diamond et al., "Rapid Evolution of Ant Thermal Tolerance across an Urban-Rural Temperature Cline," *Biological Journal of the Linnean Society* 121, no. 2 (February 21, 2017): 248–257.

17. Lacy D. Chick et al., "Urban Heat Islands Advance the Timing of Reproduction in a Social Insect," *Journal of Thermal Biology* 80 (February 2019): 119–125.

18. Lacy D. Chick et al., "Urban Heat Islands Advance the Timing of Reproduction in a Social Insect," *Journal of Thermal Biology* 80 (February 2019): 119–125.

19. Stephen E. Harris et al., "Signatures of Rapid Evolution in Urban and Rural Transcriptomes of White-Footed Mice (Peromyscus leucopus) in the New York Metropolitan Area," *PLoS One* 8, no. 8 (August 28, 2013): e74938.

20. Amélie Desvars-Larrive et al., "Population Genetics, Community of Parasites, and Resistance to Rodenticides in an Urban Brown Rat (Rattus norvegicus) Population," *PLoS One* 12, no. 9 (September 8, 2017): e0184015.

21. Atwell et al., "Boldness Behavior and Stress Physiology in a Novel Urban Environment Suggest Rapid Correlated Evolutionary Adaptation."

22. Barbara Clucas and John M. Marzluff, "Coupled Relationships between Humans and Other Organisms in Urban Areas," in *Urban Ecology* (Oxford University Press, 2011), 135–147; Mitchell Lyons et al., "Bird Interactions with Drones, from Individuals to Large Colonies," *Australian Field Ornithology* 35 (2018): 51–56.

23. Conor C. Taff and Maren N. Vitousek, "Endocrine Flexibility: Optimizing Phenotypes in a Dynamic World?" *Trends in Ecology & Evolution* 31, no. 6 (June 2016): 476–488; Claudio Carere, Doretta Caramaschi, and Tim W. Fawcett, "Covariation between Personalities and Individual Differences in Coping with Stress: Converging Evidence and Hypotheses," *Current Zoology* 56, no. 6 (December 1, 2010): 728–740.

24. Christopher J. Walsh, Allison H. Roy, Jack W. Feminella, Peter D. Cottingham, Peter M. Groffman, and Raymond P. Morgan II, "The Urban Stream Syndrome: Current Knowledge and the Search for a Cure," *Journal of the North American Benthological Society* 24, no. 3 (2005): 706–723.

25. Reid et al., "Genomic Landscape."

26. K. B. Oke, C. J. Cunningham, P. A. H. Westley, M. L. Baskett, S. M. Carlson, J. Clark, A. P. Hendry, V. A. Karatayev, N. W. Kendall, J. Kibele, H. K. Kindsvater, K. M. Kobayashi, B. Lewis, S. B. Munch, J. D. Reynolds, G. K. Vick, and E. P. Palkovacs,

"Recent Declines in Salmon Body Size Impact Ecosystems and Fisheries," *Nature Communications* 11 (2020): 4155.

27. S. M. Carlson, C. J. Cunningham, and P. A. H. Westley, "Eco-Evolutionary Dynamics in Pacific Salmon," *Heredity* 106, no. 3 (2011): 438–447.

28. Ken A. Thompson, Loren H. Rieseberg, and Dolph Schluter, "Speciation and the City," *Trends in Ecology & Evolution* 33, no. 11 (November 2018): 815–826.

29. Alberti, "Eco-Evolutionary Dynamics in an Urbanizing Planet."

30. Thompson, Rieseberg, and Schluter, "Speciation and the City"; Wouter Halfwerk, "How Should We Study Urban Speciation?" *Frontiers in Ecology and Evolution* 8 (January 11, 2021): https://doi.org/10.3389/fevo.2020.573545.

31. Patrik Nosil, Luke J. Harmon, and Ole Seehausen, "Ecological Explanations for (Incomplete) Speciation," *Trends in Ecology & Evolution* 24, no. 3 (March 2009): 145–156.

32. Marina Alberti et al., "Global Urban Signatures of Phenotypic Change in Animal and Plant Populations," *Proceedings of the National Academy of Sciences of the United States of America* 114, no. 34 (August 22, 2017): 8951–8956.

33. H. Bernard D. Kettlewell, "A Survey of the Frequencies of *Biston Betularia* (L.) (Lep.) and Its Melanic Forms in Great Britain," *Heredity* 12, no. 1 (1958): 51–72.

34. Pamela J. Yeh, "Rapid Evolution of a Sexually Selected Trait Following Population Establishment in a Novel Habitat," *Evolution; International Journal of Organic Evolution* 58, no. 1 (January 2004): 166–174.

35. Eleanor S. Diamant and Pamela J. Yeh, "Complex Patterns of Morphological Diversity across Multiple Populations of an Urban Bird Species," *Evolution; International Journal of Organic Evolution* 78, no. 7 (July 1, 2024): 1325–1337; Eleanor S. Diamant, Martin Kang, and Pamela J. Yeh, "Patterns of a Sexually Selected Trait Are Not Predictable across Urban and Non-Urban Populations of a Songbird, the Dark-Eyed Junco" (2024), https://doi.org/10.2139/ssrn.4749264.

36. Brans et al., "The Heat Is On."

37. Diane E. Nacci, Denise Champlin, and Saro Jayaraman, "Adaptation of the Estuarine Fish Fundulus Heteroclitus (Atlantic Killifish) to Polychlorinated Biphenyls (PCBs)," *Estuaries and Coasts* 33, no. 4 (July 1, 2010): 853–864; Andrew Whitehead et al., "When Evolution Is the Solution to Pollution: Key Principles, and Lessons from Rapid Repeated Adaptation of Killifish (Fundulus Heteroclitus) Populations," *Evolutionary Applications* 10, no. 8 (September 2017): 762–783.

38. Pierre-Olivier Cheptou et al., "Adaptation to Fragmentation: Evolutionary Dynamics Driven by Human Influences," *Philosophical Transactions of the Royal Society of London, Series B, Biological Sciences* 372, no. 1712 (January 19, 2017): 20160037.

39. Diamond et al., "Rapid Evolution of Ant Thermal Tolerance Across an Urban-Rural Temperature Cline."

40. Kristien I. Brans and Luc De Meester, "City Life on Fast Lanes: Urbanization Induces an Evolutionary Shift towards a Faster Lifestyle in the Water Flea Daphnia," *Functional Ecology* 32, no. 9 (2018): 2225–2240.

41. Isaac Wirgin et al., "Mechanistic Basis of Resistance to PCBs in Atlantic Tomcod from the Hudson River," *Science*, March 11, 2011.

42. Andrew Whitehead et al., "Salinity- and Population-Dependent Genome Regulatory Response during Osmotic Acclimation in the Killifish (*Fundulus heteroclitus*) Gill," *The Journal of Experimental Biology* 215, pt. 8 (April 15, 2012): 1293–1305.

43. Anne Charmantier et al., "Urbanization Is Associated with Divergence in Pace-of-Life in Great Tits," *Frontiers in Ecology and Evolution* 5 (2017): 53.

44. Pamela J. Yeh and Trevor D. Price, "Adaptive Phenotypic Plasticity and the Successful Colonization of a Novel Environment," *The American Naturalist* 164, no. 4 (October 1, 2004): 531–542.

45. Kristin M. Winchell et al., "Phenotypic Shifts in Urban Areas in the Tropical Lizard Anolis Cristatellus," *Evolution: International Journal of Organic Evolution* 70, no. 5 (May 2016): 1009–1022; Shane C. Campbell-Staton et al., "Parallel Selection on Thermal Physiology Facilitates Repeated Adaptation of City Lizards to Urban Heat Islands," *Nature Ecology & Evolution* 4, no. 4 (April 2020): 652–658; Kristin M. Winchell, Liam J. Revell, R. Graham Reynolds, and Jason J. Kolbe, "Genome-Wide Parallelism Underlies Contemporary Adaptation in Urban Lizards," *Proceedings of the National Academy of Sciences* 120, no. 13 (2023): e2216789120.

46. Palkovacs et al., "Fates Beyond Traits"; Colin M. Donihue and Max R. Lambert, "Adaptive Evolution in Urban Ecosystems," *Ambio* 44, no. 3 (April 2015): 194–203; Mark J. McDonnell and Amy K. Hahs, "Adaptation and Adaptedness of Organisms to Urban Environments," *Annual Review of Ecology, Evolution, and Systematics* 46, no. 1 (December 4, 2015): 261–280; Alberti et al., "Global Urban Signatures of Phenotypic Change in Animal and Plant Populations"; Marc T. J. Johnson and Jason Munshi-South, "Evolution of Life in Urban Environments," *Science* 358, no. 6363 (November 3, 2017): https://doi.org/10.1126/science.aam8327; L. Ruth Rivkin et al., "A Roadmap for Urban Evolutionary Ecology," *Evolutionary Applications* 12, no. 3 (2019): 384–398.

47. Stefan Leyk et al., "The Heterogeneity and Change in the Urban Structure of Metropolitan Areas in the United States, 1990–2010," *Scientific Data* 6, no. 1 (December 16, 2019): 1–15; Alberti et al., "The Complexity of Urban Eco-Evolutionary Dynamics."

48. Gary E. Machlis, Jo Ellen Force, and William R. Burch Jr., "The Human Ecosystem Part I: The Human Ecosystem as an Organizing Concept in Ecosystem Management," *Society & Natural Resources* 10, no. 4 (July 1, 1997): 347–367.

49. Steward T. A. Pickett, Mary Cadenasso, and Clive Jones, "Generation of Heterogeneity by Organisms: Creation, Maintenance, and Transformation," in *The Ecological Consequences of Environmental Heterogeneity*, ed. Michael Hutchings, Elizabeth John, and Alan Stewart (Oxford, UK: Blackwell, 2000), 33–52.

50. Brans et al., "The Heat Is On"; Brans, Stoks, and De Meester, "Urbanization Drives Genetic Differentiation in Physiology and Structures the Evolution of Pace-of-Life Syndromes in the Water Flea Daphnia Magna"; Brans and De Meester, "City Life on Fast Lanes."

51. Brans et al., "The Heat Is On."

52. Brans, Stoks, and De Meester, "Urbanization Drives Genetic Differentiation in Physiology and Structures the Evolution of Pace-of-Life Syndromes in the Water Flea Daphnia Magna."

53. Harris et al., "Signatures of Rapid Evolution in Urban and Rural Transcriptomes of White-Footed Mice (Peromyscus leucopus) in the New York Metropolitan Area"; Stephen E. Harris and Jason Munshi-South, "Signatures of Positive Selection and Local Adaptation to Urbanization in White-Footed Mice (*Peromyscus leucopus*)," *Molecular Ecology* 26, no. 22 (November 2017): 6336–6350.

54. Arbel Harpak et al., "Genetic Adaptation in New York City Rats," *Genome Biology and Evolution* 13, no. 1 (January 7, 2021): https://doi.org/10.1093/gbe/evaa247.

55. Christopher J. Schell et al., "Parental Habituation to Human Disturbance over Time Reduces Fear of Humans in Coyote Offspring," *Ecology and Evolution* 8, no. 24 (December 2018): 12965–12980.

56. Christopher J. Schell et al., "The Evolutionary Consequences of Human-Wildlife Conflict in Cities," *Evolutionary Applications* 14, no. 1 (January 2021): 178–197.

57. Christopher J. Schell et al., "The Ecological and Evolutionary Consequences of Systemic Racism in Urban Environments," *Science* 369, no. 6510 (September 18, 2020): eaay4497.

58. Eyal Shochat et al., "From Patterns to Emerging Processes in Mechanistic Urban Ecology," *Trends in Ecology & Evolution* 21, no. 4 (April 2006): 186–191.

59. Alberti et al., "The Complexity of Urban Eco-Evolutionary Dynamics"; Marta Szulkin, Jason Munshi-South, and Anne Charmantier, *Urban Evolutionary Biology* (Oxford, UK: Oxford University Press, 2020).

Chapter 2

1. Nicole L. Boivin et al., "Ecological Consequences of Human Niche Construction: Examining Long-Term Anthropogenic Shaping of Global Species Distributions,"

Proceedings of the National Academy of Sciences of the United States of America 113, no. 23 (June 7, 2016): 6388–6396.

2. Erle C. Ellis et al., "People Have Shaped Most of Terrestrial Nature for at Least 12,000 Years," *Proceedings of the National Academy of Sciences of the United States of America* 118, no. 17 (April 27, 2021): e2023483118.

3. Michael D. Purugganan and Dorian Q. Fuller, "The Nature of Selection During Plant Domestication," *Nature* 457, no. 7231 (February 12, 2009): 843–848.

4. David M. J. S. Bowman et al., "The Human Dimension of Fire Regimes on Earth," *Journal of Biogeography* 38, no. 12 (December 2011): 2223–2236.

5. Paul L. Koch and Anthony D. Barnosky, "Late Quaternary Extinctions: State of the Debate," *Annual Review of Ecology, Evolution, and Systematics* 37, no. 1 (December 1, 2006): 215–250.

6. Anthony D. Barnosky et al., "Variable Impact of Late-Quaternary Megafaunal Extinction in Causing Ecological State Shifts in North and South America," *Proceedings of the National Academy of Sciences of the United States of America* 113, no. 4 (January 26, 2016): 856–861.

7. Boivin et al., "Ecological Consequences of Human Niche Construction."

8. Alberti, "Cities of the Anthropocene."

9. Alberti, "Cities of the Anthropocene."

10. Robert H. MacArthur and Edward O. Wilson, *The Theory of Island Biogeography* (Princeton, NJ: Princeton University Press, 2001).

11. Simon T. Segar et al., "The Role of Evolution in Shaping Ecological Networks," *Trends in Ecology & Evolution* 35, no. 5 (May 2020): 454–466.

12. Sophie Cardinal and Bryan N. Danforth, "Bees Diversified in the Age of Eudicots," *Proceedings: Biological Sciences* 280, no. 1755 (March 22, 2013): 20122686.

13. Franziska S. Brunner et al., "Experimental Evidence That Parasites Drive Eco-Evolutionary Feedbacks," *Proceedings of the National Academy of Sciences of the United States of America* 114, no. 14 (April 4, 2017): 3678–3683.

14. Timothée Poisot and Daniel B. Stouffer, "How Ecological Networks Evolve," *BioRxiv* (bioRxiv, August 29, 2016), https://doi.org/10.1101/071993.

15. Segar et al., "The Role of Evolution in Shaping Ecological Networks."

16. Lindsay S. Miles et al., "Gene Flow and Genetic Drift in Urban Environments," *Molecular Ecology* 28, no. 18 (2019): 4138–4151.

17. Bullock et al., "Human-Mediated Dispersal and the Rewiring of Spatial Networks.

18. Bullock et al., "Human-Mediated Dispersal and the Rewiring of Spatial Networks."

19. María del Mar Delgado et al., "Prospecting and Dispersal: Their Eco-Evolutionary Dynamics and Implications for Population Patterns," *Proceedings: Biological Sciences* 281, no. 1778 (March 7, 2014): 20132851.

20. Lisa G. Crozier, Kristin M. McClure, Timothy D. Beechie, Steven J. Bograd, Daniel A. Boughton, David M. Carr, Heidi C. Cooley, Michael M. Huff, et al., "Snake River Sockeye and Chinook Salmon in a Changing Climate," *PLoS ONE* 15, no. 9 (2020): e0238882.

21. Jeffrey J. Duda, Christopher E. Torgersen, Scott J. Brenkman, Randall J. Peters, Patrick Crain, George R. Pess, and Michael McHenry, "Reconnecting the Elwha River: Spatial Patterns of Fish Response to Dam Removal," *Frontiers in Ecology and Evolution* 9 (2021): 765488

22. Andreas Wagner, *Robustness and Evolvability in Living Systems* (Princeton, NJ: Princeton University Press, 2013).

23. C. R. Landry, D. L. Hartl, and J. M. Ranz, "Genome Clashes in Hybrids: Insights from Gene Expression," *Heredity* 99, no. 5 (November 2007): 483–493.

24. John N. Thompson, *The Geographic Mosaic of Coevolution*, Interspecific Interactions (Chicago, IL: University of Chicago Press, 2005); Conrad Labandeira, "The Origin of Herbivory on Land: Initial Patterns of Plant Tissue Consumption by Arthropods," *Insect Science* 14, no. 4 (August 2007): 259–275.

25. Duncan J. Watts and Steven H. Strogatz, "Collective Dynamics of 'Small-World' Networks," *Nature* 393, no. 6684 (1998): 440–442.

26. Konstantinos Tzoulas et al., "Promoting Ecosystem and Human Health in Urban Areas Using Green Infrastructure: A Literature Review," *Landscape and Urban Planning* 81, no. 3 (June 20, 2007): 167–178.

27. Erik Andersson et al., "Reconnecting Cities to the Biosphere: Stewardship of Green Infrastructure and Urban Ecosystem Services," *Ambio* 43, no. 4 (May 2014): 445–453.

28. Amanda D. Rodewald and Stanley D. Gehrt, "Wildlife Population Dynamics in Urban Landscapes," in *Urban Wildlife Conservation: Theory and Practice*, ed. Robert A. McCleery, Christopher E. Moorman, and M. Nils Peterson (Boston: Springer US, 2014), 117–147.

29. Cheptou et al., "Rapid Evolution of Seed Dispersal in an Urban Environment in the Weed Crepis Sancta."

30. Robert V. O'Neill, "Is It Time to Bury the Ecosystem Concept? (With Full Military Honors, of Course!)," *Ecology* 82 (2001): 3275–3284.

31. G. Evelyn Hutchinson, *The Ecological Theater and the Evolutionary Play* (New Haven, CT: Yale University Press, 1965).

32. Will Steffen et al., "The Anthropocene: Conceptual and Historical Perspectives," *Philosophical Transactions. Series A, Mathematical, Physical, and Engineering Sciences* 369, no. 1938 (March 13, 2011): 842–867; Will Steffen et al., "The Trajectory of the Anthropocene: The Great Acceleration," *The Anthropocene Review* 2, no. 1 (April 2015): 81–98.

33. Jaia Syvitski et al., "Author Correction: Extraordinary Human Energy Consumption and Resultant Geological Impacts Beginning around 1950 CE Initiated the Proposed Anthropocene Epoch," *Communications Earth & Environment* 1, no. 1 (November 2, 2020), https://doi.org/10.1038/s43247-020-00049-8.

34. Kevin N. Laland, *Darwin's Unfinished Symphony* (Princeton, NJ: Princeton University Press, 2018).

35. Peter J. Richerson and Robert Boyd, "The Human Life History Is Adapted to Exploit the Adaptive Advantages of Culture," *Philosophical Transactions of the Royal Society of London, Series B, Biological Sciences* 375, no. 1803 (July 20, 2020): 20190498.

36. Kim R. Hill et al., "Co-Residence Patterns in Hunter-Gatherer Societies Show Unique Human Social Structure," *Science (New York, N.Y.)* 331, no. 6022 (March 11, 2011): 1286–1289.

Chapter 3

1. Lynn Sagan, "On the Origin of Mitosing Cells," *Journal of Theoretical Biology* 14, no. 3 (March 1967): 255–274; Robert M. Schwartz and Margaret O. Dayhoff, "Origins of Prokaryotes, Eukaryotes, Mitochondria, and Chloroplasts," *Science* 199, no. 4327 (January 27, 1978): 395–403; Lynn Margulis, "Symbiogenesis and Symbionticism," in *Symbiosis as a Source of Evolutionary Innovation*, ed. Lynn Margulis and René Fester (Cambridge, MA: MIT Press, 1991), 1–13.

2. Thomas C. Scott-Phillips et al., "The Niche Construction Perspective: A Critical Appraisal," *Evolution: International Journal of Organic Evolution* 68, no. 5 (May 2014): 1231–1243.

3. Clive G. Jones, John H. Lawton, and Moshe Shachak, "Organisms as Ecosystem Engineers," in *Ecosystem Management* (New York: Springer, 1994), 130–147; F. John Odling-Smee, Kevin N. Laland, and Marcus W. Feldman, *Niche Construction: The Neglected Process in Evolution* (Princeton, NJ: Princeton University Press, 2003); F. John Odling-Smee et al., "Niche Construction Theory: A Practical Guide for Ecologists," *The Quarterly Review of Biology* 88, no. 1 (2013): 3–28; Sonia E. Sultan, *Organism and Environment: Ecological Development, Niche Construction, and Adaptation* (Oxford: Oxford University Press, 2015).

4. Fanie Pelletier, Dany Garant, and Andrew P. Hendry, "Eco-Evolutionary Dynamics," *Philosophical Transactions of the Royal Society of London, Series B, Biological Sciences* 364, no. 1523 (June 12, 2009): 1483–1489; Post and Palkovacs, "Eco-Evolutionary Feedbacks in Community and Ecosystem Ecology"; Andrew P. Hendry, *Eco-Evolutionary Dynamics* (Princeton, NJ: Princeton University Press, 2016); Pimentel, "Animal Population Regulation by the Genetic Feedback Mechanism."

5. Blake Matthews et al., "Under Niche Construction: An Operational Bridge Between Ecology, Evolution, and Ecosystem Science," *Ecological Monographs* 84, no. 2 (May 2014): 245–263.

6. Richard C. Lewontin, "Gene, Organism and Environment," *Evolution from Molecules to Men* 273 (1983): 285.

7. Lewontin, "Gene, Organism and Environment," 276. Emphasis added.

8. F. John Odling-Smee, "Niche-Constructing Phenotypes," in *The Role of Behavior in Evolution,* ed. Henry C. Plotkin, vol. 198 (Cambridge, MA: MIT Press, 1988), 73–132.

9. Odling-Smee, Laland, and Feldman, *Niche Construction.*

10. Kevin Laland, John Odling-Smee, and John Endler, "Niche Construction, Sources of Selection and Trait Coevolution," *Interface Focus* 7, no. 5 (October 6, 2017): 20160147.

11. Andrew D. Clark et al., "Niche Construction Affects the Variability and Strength of Natural Selection," *The American Naturalist* 195, no. 1 (January 2020): 16–30.

12. Ingeborg Haug, Sabrina Setaro, and Juan Pablo Suárez, "Reforestation Sites Show Similar and Nested AMF Communities to an Adjacent Pristine Forest in a Tropical Mountain Area of South Ecuador," *PLoS One* 8, no. 5 (May 6, 2013): e63524.

13. Robert D. Holt, "Linking Species and Ecosystems: Where's Darwin?" in *Linking Species & Ecosystems,* ed. C. G. Jones and J. H. Lawton (Boston: Springer US, 1995), 273–279; Blake Matthews et al., "Evolution as an Ecosystem Process: Insights from Genomics," *Genome* 61, no. 4 (April 2018): 298–309; Hendry, *Eco-Evolutionary Dynamics.*

14. Erik F. Y. Hom and Alexandra S. Penn, "Symbiosis and the Anthropocene," *Symbiosis* 84, no. 3 (September 3, 2021): 239–270.

15. Hom and Penn, "Symbiosis and the Anthropocene," 240.

16. Timon McPhearson et al., "Advancing Urban Ecology Toward a Science of Cities," *Bioscience* 66, no. 3 (March 1, 2016): 198–212; Amy K. Hahs et al., "A Global Synthesis of Plant Extinction Rates in Urban Areas," *Ecology Letters* 12, no. 11 (2009): 1165–1173.

17. Stanley H. Faeth, Christofer Bang, and Susanna Saari, "Urban Biodiversity: Patterns and Mechanisms," *Annals of the New York Academy of Sciences* 1223 (March 2011): 69–81.

18. Philip E. Hulme, "Trade, Transport and Trouble: Managing Invasive Species Pathways in an Era of Globalization," *The Journal of Applied Ecology*, 2009; Anthony Ricciardi et al., "Invasion Science: A Horizon Scan of Emerging Challenges and Opportunities," *Trends in Ecology & Evolution* 32, no. 6 (June 2017): 464–474.

19. Myla F. J. Aronson et al., "A Global Analysis of the Impacts of Urbanization on Bird and Plant Diversity Reveals Key Anthropogenic Drivers," *Proceedings of the Royal Society B: Biological Sciences* 281, no. 1780 (April 7, 2014): 20133330; Ingo Kowarik, "Novel Urban Ecosystems, Biodiversity, and Conservation," *Environmental Pollution* 159, nos. 8–9 (August 2011): 1974–1983.

20. John Vandermeer et al., "Increased Competition May Promote Species Coexistence," *Proceedings of the National Academy of Sciences of the United States of America* 99, no. 13 (June 25, 2002): 8731–8736.

21. Willem Stock et al., "Human Impact on Symbioses Between Aquatic Organisms and Microbes," *Aquatic Microbial Ecology: International Journal* 87 (August 26, 2021): 113–138; M. Denise Dearing, Martin Kaltenpoth, and Jonathan Gershenzon, "Demonstrating the Role of Symbionts in Mediating Detoxification in Herbivores," *Symbiosis* 87, no. 1 (September 13, 2022): 59–66.

22. Hom and Penn, "Symbiosis and the Anthropocene."

23. Erik F. Y. Hom and Andrew W. Murray, "Plant-Fungal Ecology. Niche Engineering Demonstrates a Latent Capacity for Fungal-Algal Mutualism," *Science* 345, no. 6192 (July 4, 2014): 94–98.

Chapter 4

1. C. S. Holling, "Resilience and Stability of Ecological Systems," *Annual Review of Ecology and Systematics* 4, no. 1 (November 1973): 1–23.

2. Alberti et al., "Evolving Nature-Based Solutions."

3. Crawford S. Holling, "Resilience and Stability of Ecological Systems," *Annual Review of Ecology and Systematics* 4, no. 1 (1973): 1–23.

4. Simon A. Levin et al., "Governance in the Face of Extreme Events: Lessons from Evolutionary Processes for Structuring Interventions, and the Need to Go Beyond," *Ecosystems* 25, no. 3 (2022): 697–711.

5. Jonathan L. Payne et al., "The Evolution of Complex Life and the Stabilization of the Earth System," *Interface Focus* 10, no. 4 (August 6, 2020): 20190106.

6. Christopher Blake, Jake N. Barber, Tim Connallon, and Michael J. McDonald, "Evolutionary Shift of a Tipping Point Can Precipitate, or Forestall, Collapse in a

Microbial Community," *Nature Ecology & Evolution* (September 18, 2024), https://doi.org/10.1038/s41559-024-02543-0.

7. Yuebo Su et al., "Leaf Functional Traits Vary in Urban Environments: Influences of Leaf Age, Land-Use Type, and Urban–Rural Gradient," *Frontiers in Ecology and Evolution* 9 (2021): 681959.

8. Frank Ian Woodward and Colleen K. Kelly, "The Influence of CO_2 Concentration on Stomatal Density," *The New Phytologist* 131, no. 3 (November 1995): 311–327.

9. William K. Smith, "Temperatures of Desert Plants: Another Perspective on the Adaptability of Leaf Size," *Science* 201, no. 4356 (August 18, 1978): 614–616.

10. P. B. Reich et al., "The Evolution of Plant Functional Variation: Traits, Spectra, and Strategies," *International Journal of Plant Sciences* 164, no. S3 (May 2003): S143–S164; P. B. Reich, M. B. Walters, and D. S. Ellsworth, "From Tropics to Tundra: Global Convergence in Plant Functioning," *Proceedings of the National Academy of Sciences of the United States of America* 94, no. 25 (December 9, 1997): 13730–13734.

11. I. J. Wright, P. B. Reich, and M. Westoby, "Strategy Shifts in Leaf Physiology, Structure and Nutrient Content between Species of High- and Low-rainfall and High- and Low-nutrient Habitats," *Functional Ecology* 15, no. 4 (August 2001): 423–434.

12. Su et al., "Leaf Functional Traits Vary in Urban Environments."

13. Lei Wang et al., "Physicochemical Characteristics of Ambient Particles Settling upon Leaf Surfaces of Urban Plants in Beijing," *Journal of Environmental Sciences* 18, no. 5 (September 1, 2006): 921–926.

14. Andrew P. Hendry et al., "The Contemporary Evolution of Fitness," *Annual Review of Ecology, Evolution, and Systematics* 49 (2018): 457–476.

15. Stanley H. Faeth et al., "Trophic Dynamics in Urban Communities," *Bioscience* 55, no. 5 (May 1, 2005): 399–407; Rodewald and Gehrt, "Wildlife Population Dynamics in Urban Landscapes."

16. Vincent Fugère and Andrew P. Hendry, "Human Influences on the Strength of Phenotypic Selection," *Proceedings of the National Academy of Sciences of the United States of America* 115, no. 40 (October 2, 2018): 10070–10075.

17. Rose Thorogood et al., "Understanding and Applying Biological Resilience, from Genes to Ecosystems," *NPJ Biodiversity* 2, no. 1 (August 28, 2023): 1–13.

18. Thomas Suddendorf and Michael C. Corballis, "The Evolution of Foresight: What Is Mental Time Travel, and Is It Unique to Humans?" *The Behavioral and Brain Sciences* 30, no. 3 (June 2007): 299–313; discussion 313–531.

19. Clint J. Perry and Lars Chittka, "How Foresight Might Support the Behavioral Flexibility of Arthropods," *Current Opinion in Neurobiology* 54 (February 2019): 171–177.

20. Chongxi Lai et al., "Volitional Activation of Remote Place Representations with a Hippocampal Brain-Machine Interface," *Science* 382, no. 6670 (November 3, 2023): 566–573.

21. François Huber, *Nouvelles observations sur les abeilles / New Observations Upon Bees* (Geneva: J. J. Paschoud, 1814).

22. Perry and Chittka, "How Foresight Might Support the Behavioral Flexibility of Arthropods."

23. Peter J. Richerson and Robert Boyd, *Not by Genes Alone: How Culture Transformed Human Evolution* (Chicago: University of Chicago Press, 2005).

24. Carrie Hritz et al., *If the Past Teaches, What Does the Future Learn? Ancient Urban Regions and the Durable Future*, ed. John T. Murphy and Carole L. Crumley (Delft, Netherlands: TU Delft, 2022).

25. Eyal Shochat et al., "Invasion, Competition, and Biodiversity Loss in Urban Ecosystems," *Bioscience* 60, no. 3 (March 1, 2010): 199–208; Mark J. McDonnell and Amy K. Hahs, "The Future of Urban Biodiversity Research: Moving Beyond the 'Low-Hanging Fruit,'" *Urban Ecosystems* 16, no. 3 (September 2013): 397–409; McDonnell and Hahs, "Adaptation and Adaptedness of Organisms to Urban Environments"; Myla F. J. Aronson et al., "Hierarchical Filters Determine Community Assembly of Urban Species Pools," *Ecology* 97, no. 11 (November 2016): 2952–2963; Tibor Magura, Eszter Kiss, and Gábor L. Lövei, "No Consistent Diversity Patterns in Terrestrial Mammal Assemblages Along Rural-Urban Forest Gradients," *Basic and Applied Ecology* 52 (May 1, 2021): 38–45.

26. Miles et al., "Gene Flow and Genetic Drift in Urban Environments"; Abe Perez and Sarah E. Diamond, "Idiosyncrasies in Cities: Evaluating Patterns and Drivers of Ant Biodiversity Along Urbanization Gradients," *Journal of Urban Ecology* 5, no. 1 (2019): juz01; Chloé Schmidt et al., "Continent-Wide Effects of Urbanization on Bird and Mammal Genetic Diversity," *Proceedings: Biological Sciences* 287, no. 1920 (February 12, 2020): 20192497.

27. Marina Alberti and Tianzhe Wang, "Detecting Patterns of Vertebrate Biodiversity Across the Multidimensional Urban Landscape," *Ecology Letters* 25 (April 1, 2022): 1027–1045.

28. A. Mougi and M. Kondoh, "Diversity of Interaction Types and Ecological Community Stability," *Science* 337, no. 6092 (July 20, 2012): 349–351; Marc Peipoch et al., "Ecological Simplification: Human Influences on Riverscape Complexity," *Bioscience* 65, no. 11 (November 1, 2015): 1057–1065.

29. Robert V. O'Neill, Robert H. Gardner, and Donald E. Weller, "Chaotic Models as Representations of Ecological Systems," *The American Naturalist* 120, no. 2 (August 1, 1982): 259–263; O'Neill, "Is It Time to Bury the Ecosystem Concept? (With Full Military Honors, of Course!)"; Steward T. A. Pickett, V. Thomas Parker, and Peggy L.

Fiedler, "The New Paradigm in Ecology: Implications for Conservation Biology Above the Species Level," in *Conservation Biology: The Theory and Practice of Nature Conservation Preservation and Management*, ed. Peggy L. Fiedler and Subodh K. Jain (Boston: Springer US, 1992), 65–88; Simon A. Levin, "Towards a Science of Ecological Management," *Conservation Ecology* 3, no. 2 (1999), http://www.consecol.org/vol3/iss2/art6/.

30. Jack Ahern, "From Fail-Safe to Safe-to-Fail: Sustainability and Resilience in the New Urban World," *Landscape and Urban Planning* 100, no. 4 (April 30, 2011): 341–343.

31. Holling, "Resilience and Stability of Ecological Systems"; Crawford Stanley Holling, "Engineering Resilience Versus Ecological Resilience," *Engineering Within Ecological Constraints* 31, no. 1996 (1996): 32.

32. Holling, "Resilience and Stability of Ecological Systems."

33. Holling, "Resilience and Stability of Ecological Systems."

34. Alberti, "Cities of the Anthropocene"; Alberti et al., "Evolving Nature-Based Solutions."

35. Mahsa Fardisi et al., "Rapid Evolutionary Responses to Insecticide Resistance Management Interventions by the German Cockroach (Blattella Germanica L.)," *Scientific Reports* 9, no. 1 (June 5, 2019): 8292.

36. Jose Miguel Sandoval-Gil, Juan M. Ruiz, and Lázaro Marín-Guirao, "Advances in Understanding Multilevel Responses of Seagrasses to Hypersalinity," *Marine Environmental Research* 183, no. 105809 (January 2023): 105809.

37. P. B. Reich et al., "The Evolution of Plant Functional Variation: Traits, Spectra, and Strategies," *International Journal of Plant Sciences* 164, no. S3 (May 2003): S143–S164.

38. Alberti, "Cities of the Anthropocene."

39. Alberti et al., "Evolving Nature-Based Solutions."

40. Johanna Andrea Martínez-Villa, Alain Paquette, Kenneth J. Feeley, Paula Andrea Morales-Morales, Christian Messier, and Sandra M. Durán, "Changes in Morphological and Physiological Traits of Urban Trees in Response to Elevated Temperatures within an Urban Heat Island," *Tree Physiology* 44, no. 12 (December 2024): tpae145.

41. Panagiotis Theodorou et al., "Urban Areas as Hotspots for Bees and Pollination but Not a Panacea for All Insects," *Nature Communications* 11, no. 1 (January 29, 2020): 576.

42. M. L. Vahsen et al., "Rapid Plant Trait Evolution Can Alter Coastal Wetland Resilience to Sea Level Rise," *Science (New York, N.Y.)* 379, no. 6630 (January 27, 2023): 393–398.

43. Brans, Stoks, and De Meester, "Urbanization Drives Genetic Differentiation in Physiology and Structures the Evolution of Pace-of-Life Syndromes in the Water Flea Daphnia Magna."

44. Nelson G. Hairston et al., "Rapid Evolution Revealed by Dormant Eggs," *Nature* 401, no. 6752 (September 1999): 446; Nelson G. Hairston et al., "Natural Selection for Grazer Resistance to Toxic Cyanobacteria: Evolution of Phenotypic Plasticity?" *Evolution: International Journal of Organic Evolution* 55, no. 11 (2001): 2203–2214.

45. Robert S. Smith et al., "Oyster Restoration to Recover Ecosystem Services," *Annual Review of Marine Science* 17 (January 2025).

46. Agnese Lupo, Marisa Haenni, and Jean-Yves Madec, "Antimicrobial Resistance in *Acinetobacter* spp. and *Pseudomonas* spp.," *Microbiology Spectrum 6,* no. 3 (June 2018): ARBA-0007-2017.

47. Alberti et al., "Evolving Nature-Based Solutions."

48. Alberti, *Cities That Think Like Planets.*

49. Carl Folke et al., "Social-Ecological Resilience and Biosphere-Based Sustainability Science," *Ecology and Society* 21, no. 3 (2016): 41.

50. Craig R. Allen and Crawford Stanley Holling, "Novelty, Adaptive Capacity, and Resilience," *Ecology and Society* 15, no. 3 (2010): 24; Pietro Landi et al., "Complexity and Stability of Ecological Networks: A Review of the Theory," *Population Ecology* 60, no. 4 (October 1, 2018): 319–345; David Houle and Daniela M. Rossoni, "Complexity, Evolvability, and the Process of Adaptation," *Annual Review of Ecology, Evolution, and Systematics* 53, no. 1 (November 2, 2022): 137–159.

51. Marten Scheffer et al., "Anticipating Critical Transitions," *Science* 338, no. 6105 (October 19, 2012): 344–348.

52. Alberti, "Cities of the Anthropocene"; Carla M. Sgrò, Andrew J. Lowe, and Ary A. Hoffmann, "Building Evolutionary Resilience for Conserving Biodiversity Under Climate Change," *Evolutionary Applications* 4, no. 2 (March 2011): 326–337; Thomas B. Smith et al., "Prescriptive Evolution to Conserve and Manage Biodiversity," *Annual Review of Ecology, Evolution, and Systematics* 45, no. 1 (2014): 1–22; Scott P. Carroll et al., "Applying Evolutionary Biology to Address Global Challenges," *Science* 346, no. 6207 (October 17, 2014): 1245993; Peter Søgaard Jørgensen, Carl Folke, and Scott P. Carroll, "Evolution in the Anthropocene: Informing Governance and Policy," *Annual Review of Ecology, Evolution, and Systematics* 50, no. 1 (November 2, 2019): 527–546.

53. Alberti, "Cities of the Anthropocene."

54. Erica N. Spotswood et al., "The Biological Deserts Fallacy: Cities in Their Landscapes Contribute More Than We Think to Regional Biodiversity," *Bioscience* 71, no. 2 (February 2021): 148–160.

55. Laura M. Thompson et al., "Connecting Research and Practice to Enhance the Evolutionary Potential of Species Under Climate Change," *Conservation Science and Practice,* January 3, 2023.

56. Su et al., "Leaf Functional Traits Vary in Urban Environments."

57. Jason Munshi-South, Christine P. Zolnik, and Stephen E. Harris, "Population Genomics of the Anthropocene: Urbanization Is Negatively Associated with Genome-Wide Variation in White-Footed Mouse Populations," *Evolutionary Applications* 9, no. 4 (2016): 546–564.

58. Christopher A. Lepczyk, Myla F. J. Aronson, and Frank A. La Sorte, "Cities as Sanctuaries," *Frontiers in Ecology and the Environment*, May 11, 2023.

59. Myla F. J. Aronson et al., "Biodiversity in the City: Key Challenges for Urban Green Space Management," *Frontiers in Ecology and the Environment* 15, no. 4 (May 2017): 189–196.

60. William E. Peterman et al., "Assessing Modularity in Genetic Networks to Manage Spatially Structured Metapopulations," *Ecosphere* 7, no. 2 (February 2016): e01231; Sven Eggimann, "The Potential of Implementing Superblocks for Multifunctional Street Use in Cities," *Nature Sustainability* 5, no. 5 (May 2022): 406–414.

61. Dirk M. Lorenz, Alice Jeng, and Michael W. Deem, "The Emergence of Modularity in Biological Systems," *Physics of Life Reviews* 8, no. 2 (June 2011): 129–160; Scheffer et al., "Anticipating Critical Transitions"; Jacopo Grilli, Tim Rogers, and Stefano Allesina, "Modularity and Stability in Ecological Communities," *Nature Communications* 7 (June 23, 2016): 12031.

62. Robert J. Fletcher Jr. et al., "Network Modularity Reveals Critical Scales for Connectivity in Ecology and Evolution," *Nature Communications* 4 (2013): 2572; Agnes S. Dellinger et al., "Modularity Increases Rate of Floral Evolution and Adaptive Success for Functionally Specialized Pollination Systems," *Communications Biology* 2 (December 5, 2019): 453.

63. Garry Peterson, Craig R. Allen, and C. S. Holling, "Ecological Resilience, Biodiversity, and Scale," *Ecosystems* 1, no. 1 (January 1, 1998): 6–18; Allen and Holling, "Novelty, Adaptive Capacity, and Resilience"; Annika T. H. Keeley et al., "Governing Ecological Connectivity in Cross-Scale Dependent Systems," *Bioscience* 72, no. 4 (April 2022): 372–386.

64. Scheffer et al., "Anticipating Critical Transitions"; Vasilis Dakos and Jordi Bascompte, "Critical Slowing Down as Early Warning for the Onset of Collapse in Mutualistic Communities," *Proceedings of the National Academy of Sciences of the United States of America* 111, no. 49 (December 9, 2014): 17546–17551; Sarah P. Flanagan et al., "Guidelines for Planning Genomic Assessment and Monitoring of Locally Adaptive Variation to Inform Species Conservation," *Evolutionary Applications* 11, no. 7 (August 2018): 1035–1052.

65. Simon A. Levin, "Self-Organization and the Emergence of Complexity in Ecological Systems," *Bioscience* 55, no. 12 (December 1, 2005): 1075–1079; Allen and

Holling, "Novelty, Adaptive Capacity, and Resilience"; Li-Xia Zhao et al., "The Shaping Role of Self-Organization: Linking Vegetation Patterning, Plant Traits and Ecosystem Functioning," *Proceedings: Biological Sciences* 286, no. 1900 (April 10, 2019): 20182859.

Chapter 5

1. Lynn Margulis, *Symbiosis in Cell Evolution: Life and Its Environment on the Early Earth* (New York: W. H. Freeman & Company, 1981); Lynn Margulis, "Archaeal-Eubacterial Mergers in the Origin of Eukarya: Phylogenetic Classification of Life," *Proceedings of the National Academy of Sciences of the United States of America* 93, no. 3 (February 6, 1996): 1071–1076.

2. Timothy Beatley, *Biophilic Cities: Integrating Nature into Urban Design and Planning* (Washington, DC: Island Press, 2011); Timothy Beatley, *Handbook of Biophilic City Planning & Design* (Washington, DC: Island Press, 2016).

3. Chrisna du Plessis, "Towards a Regenerative Paradigm for the Built Environment," *Building Research and Information* 40, no. 1 (January 1, 2012): 7–22; Raymond J. Cole, "Transitioning from Green to Regenerative Design," *Building Research and Information* 40, no. 1 (January 1, 2012): 39–53.

4. Simon P. Hart, Martin M. Turcotte, and Jonathan M. Levine, "Effects of Rapid Evolution on Species Coexistence," *Proceedings of the National Academy of Sciences of the United States of America* 116, no. 6 (February 5, 2019): 2112–2117; Robert D. Holt, "Bringing the Hutchinsonian Niche into the 21st Century: Ecological and Evolutionary Perspectives," *Proceedings of the National Academy of Sciences of the United States of America* 106, suppl. 2 (November 17, 2009): 19659–19665; Mathew A. Leibold, "Ecology: Return of the Niche," *Nature* 454, no. 7200 (July 3, 2008): 39–41.

5. Alberti et al., "The Complexity of Urban Eco-Evolutionary Dynamics."

6. Nathalie Seddon, "Harnessing the Potential of Nature-Based Solutions for Mitigating and Adapting to Climate Change," *Science (New York, N.Y.)* 376, no. 6600 (June 24, 2022): 1410–1416; IUCN, "Global Standard for Nature-Based Solutions" (Gland, Switzerland: IUCN, 2020). ; Sarah E. Hobbie and Nancy B. Grimm, "Nature-Based Approaches to Managing Climate Change Impacts in Cities," *Philosophical Transactions of the Royal Society of London, Series B, Biological Sciences* 375, no. 1794 (March 16, 2020): 20190124.

7. Alberti, "Cities of the Anthropocene"; Alberti et al., "Evolving Nature-Based Solutions."

8. Brans, Stoks, and De Meester, "Urbanization Drives Genetic Differentiation in Physiology and Structures the Evolution of Pace-of-Life Syndromes in the Water Flea

Daphnia Magna"; Miner et al., "Linking Genes to Communities and Ecosystems"; Coldsnow et al., "Rapid Evolution of Tolerance to Road Salt in Zooplankton."

9. Danielle Celermajer et al., "Multispecies Justice: Theories, Challenges, and a Research Agenda for Environmental Politics," in *Trajectories in Environmental Politics* (London: Routledge, 2022), 116–137.

10. Hira Sheikh, Peta Mitchell, and Marcus Foth, "Reparative Futures of Smart Urban Governance: A Speculative Design Approach for Multispecies Justice," *Futures* 154, no. 103266 (December 2023): 103266.

11. "Kirschner-Gerhart-1998-Evolvability," n.d., https://doi.org/10.1073/pnas.95.15.8420.

12. Patrick Bateson, "Adaptability and Evolution," *Interface Focus* 7, no. 5 (October 6, 2017): 20160126.

Chapter 6

1. Bruno Glaser et al., "Carbon and Nitrogen Mineralization in Cultivated and Natural Savanna Soils of Northern Tanzania," *Biology and Fertility of Soils* 33, no. 4 (April 9, 2001): 301–309.

2. J. Lehmann, "Terra Preta Nova—Where to from Here?" in *Amazonian Dark Earths: Wim Sombroek's Vision,* ed. W. I. Woods, W. G. Teixeira, J. Lehmann, C. Steiner, A. WinklerPrins, and L. Rebellato (Dordrecht: Springer Netherlands, 2009), 473–486.

3. Charles R. Clement, "1492 and the Loss of Amazonian Crop Genetic Resources. Ii. Crop Biogeography at Contact," *Economic Botany* 53, no. 2 (April 1999): 203–216.

4. N. A. Levis, A. Serrato-Capuchina, and D. W. Pfennig, "Genetic Accommodation in the Wild: Evolution of Gene Expression Plasticity During Character Displacement," *Journal of Evolutionary Biology* 30, no. 9 (September 2017): 1712–1723.

5. Michiel Breugel et al., "Soil Nutrients and Dispersal Limitation Shape Compositional Variation in Secondary Tropical Forests Across Multiple Scales," *Journal of Ecology* 107, no. 2 (March 2019): 566–581.

6. Lisa Yeomans et al., "Waterfowl Eggshell Refines Palaeoenvironmental Reconstruction and Supports Multi-Species Niche Construction at the Pleistocene-Holocene Transition in the Levant," *Journal of Archaeological Method and Theory* 31, no. 3 (September 2024): 1383–1429.

7. Lisa Yeomans and Camilla Mazzucato, "Human-Bird Interactions in the Levant During the Late Pleistocene to Early Holocene: Multi-Scalar Analysis of Avifaunal Remains," *Archaeological and Anthropological Sciences* 16, no. 11 (November 2024), https://doi.org/10.1007/s12520-024-02090-6.

8. Hritz et al., *If the Past Teaches, What Does the Future Learn?*

9. Vernon L. Scarborough, "How to Interpret an Ancient Landscape," *Proceedings of the National Academy of Sciences of the United States of America* 100, no. 8 (April 15, 2003): 4366–4368.

10. Guillermo Algaze, *Ancient Mesopotamia at the Dawn of Civilization* (Chicago: University of Chicago Press, 2014).

11. "Human Impact on Ancient Environments. Charles L. Redman. 1999. University," n.d., https://ethnobiology.org/sites/default/files/pdfs/JoE/21-1/BRHumanImpact.pdf.

12. Ian Hodder, *Çatalhöyük: The Leopard's Tale: Revealing the Mysteries of Turkey's Ancient "Town"* (London: Thames & Hudson, 2006); Ian Kuijt, "People and Space in Early Agricultural Villages: Exploring Daily Lives, Community Size, and Architecture in the Late Pre-Pottery Neolithic," *Journal of Anthropological Archaeology* 19, no. 1 (March 2000): 75–102.

13. Chris Scarre, *Human Past: World Prehistory and Development of Human Societies* (London: Thames & Hudson, 2005).

14. Alberti, "Cities of the Anthropocene."

15. Laurel Fogarty and Nicole Creanza, "The Niche Construction of Cultural Complexity: Interactions Between Innovations, Population Size and the Environment," *Philosophical Transactions of the Royal Society of London, Series B, Biological Sciences* 372, no. 1735 (December 5, 2017): 20160428.

16. Ben Fitzhugh et al., "Human Ecodynamics: A Perspective for the Study of Long-Term Change in Socioecological Systems," *Journal of Archaeological Science, Reports* 23 (February 2019): 1077–1094.

17. Erle C. Ellis, "Ecology in an Anthropogenic Biosphere," *Ecological Monographs* 85, no. 3 (August 2015): 287–331.

18. Timothy M. Lenton and Marten Scheffer, "Spread of the Cycles: A Feedback Perspective on the Anthropocene," *Philosophical Transactions of the Royal Society of London, Series B, Biological Sciences* 379, no. 1893 (January 2024): 20220254.

19. Jason W. Moore, *Capitalism in the Web of Life* (London: Verso Books, 2015); Immanuel Wallerstein, *The Modern World System: Capitalist Agriculture and the Origins of the European World-Economy in the Sixteenth Century*, Vol 1, *Studies in Social Discontinuity* (New York: Academic Press, 1974).

20. Bruno Latour, *We Have Never Been Modern*, trans. Catherine Porter (Cambridge, MA: Harvard University Press, 1993).

21. Carolyn Merchant, "The Death of Nature," in *Science and Nature* (New York: Routledge, 2017), 12–29.

22. Moore, *Capitalism in the Web of Life.*

23. Paul J. Crutzen and Eugene F. Stoermer, "The 'Anthropocene' (2000)," in *Paul J. Crutzen and the Anthropocene: A New Epoch in Earth's History*, ed. Paul J. Crutzen (Cham: Springer International, 2021), 19–21.

24. Andrew M. Bauer et al., "Anthropocene: Event or Epoch?" *Nature* 597, no. 7876 (September 2021): 332.

25. Philip Gibbard et al., "The Anthropocene as an Event, Not an Epoch," *Journal of Quaternary Science* 37, no. 3 (April 2022): 395–399.

Chapter 7

1. Alberti, "Cities of the Anthropocene."

2. Camila I. Donatti et al., "Ensuring That Nature-Based Solutions for Climate Mitigation Address Multiple Global Challenges," *One Earth (Cambridge, Mass.)* 5, no. 5 (May 2022): 493–504; Hobbie and Grimm, "Nature-Based Approaches to Managing Climate Change Impacts in Cities."

3. Alberti, "Cities of the Anthropocene"; Alberti et al., "Evolving Nature-Based Solutions."

4. Seddon, "Harnessing the Potential of Nature-Based Solutions for Mitigating and Adapting to Climate Change."

5. John Carey, "News Feature: The Complex Case of Chesapeake Bay Restoration," *Proceedings of the National Academy of Sciences of the United States of America* 118, no. 25 (June 22, 2021): e2108734118.

6. Gourvendu Saxena et al., "Ecogenomics Reveals Metals and Land-Use Pressures on Microbial Communities in the Waterways of a Megacity," *Environmental Science & Technology* 49, no. 3 (February 3, 2015): 1462–1471.

7. Siavash Atashgahi et al., "Prospects for Harnessing Biocide Resistance for Bioremediation and Detoxification," *Science (New York, N.Y.)* 360, no. 6390 (May 18, 2018): 743–746.

8. Jiayu Liu et al., "Potential Bacterial Resources for Bioremediation of Organochlorine Pesticides and Flame Retardants Recognized from Forest Soil Across China," *Journal of Hazardous Materials*, no. 137027 (December 2024): 137027.

9. Ayad Almakki et al., "Antibiotic Resistance in Urban Runoff," *The Science of the Total Environment* 667 (June 1, 2019): 64–76; Alberto Vassallo et al., "The Bacterial Urban Resistome: Recent Advances," *Antibiotics (Basel, Switzerland)* 11, no. 4 (April 12, 2022): 512.

10. Vassallo et al., "The Bacterial Urban Resistome."

Chapter 8

1. Sheikh, Mitchell, and Foth, "Reparative Futures of Smart Urban Governance"; Petra Tschakert et al., "Multispecies Justice: Climate-Just Futures with, for and Beyond Humans," *Wiley Interdisciplinary Reviews: Climate Change* 12, no. 2 (March 2021), https://doi.org/10.1002/wcc.699; Yoshinori Nakagawa, Keishiro Hara, and Tatsuyoshi Saijo, "Becoming Sympathetic to the Needs of Future Generations: A Phenomenological Study of Participation in Future Design Workshops," Social Design Engineering Series SDES-2017-4 (Kochi University of Technology, May 19, 2017).

2. Serena Pollastri et al., "More-Than-Human Future Cities: From the Design of Nature to Designing for and Through Nature," in *Media Architecture Biennale MAB20* (unknown, 2021), https://doi.org/10.1145/3469410.3469413.

3. Donna Houston et al., "Make Kin, Not Cities! Multispecies Entanglements and 'Becoming-World' in Planning Theory," *Planning Theory* 17, no. 2 (May 1, 2018): 190–212.

4. Jennifer Wolch, Stephanie Pincetl, and Laura Pulido, "Urban Nature and the Nature of Urbanism," in *From Chicago to L.A.: Making Sense of Urban Theory*, ed. Michael J. Dear with J. Dallas Dishman (Thousand Oaks, CA: SAGE, 2002), 369–404; Jonathan Metzger, "Cultivating Torment," *City* 20, no. 4 (July 3, 2016): 581–601.

5. Jonathan Metzger, "A More-Than-Human Approach to Environmental Planning," in *The Routledge Companion to Environmental Planning* (Abingdon: Routledge, 2019), 190–99; Houston et al., "Make Kin, Not Cities!"

6. Christopher J. Schell et al., "Anthropogenic and Physiologically Induced Stress Responses in Captive Coyotes," *Journal of Mammalogy* 94, no. 5 (October 15, 2013): 1131–1140; Katie E. Robertson et al., "Stress in the City? Coyote Hair Cortisol Varies with Intrinsic and Extrinsic Factors Within a Heavily Urbanized Landscape," *The Science of the Total Environment* 901, no. 165965 (November 25, 2023): 165965; Stewart W. Breck et al., "The Intrepid Urban Coyote: A Comparison of Bold and Exploratory Behavior in Coyotes from Urban and Rural Environments," *Scientific Reports* 9, no. 1 (February 14, 2019): 2104.

7. Stanley D. Gehrt et al., "Is the Urban Coyote a Misanthropic Synanthrope? The Case from Chicago," *Cities and the Environment* 4, no. 1 (2011): 1–25.

8. Alexandra L. DeCandia et al., "Genetics of Urban Colonization: Neutral and Adaptive Variation in Coyotes (*Canis Latrans*) Inhabiting the New York Metropolitan Area," *Journal of Urban Ecology* 5, no. 1 (January 1, 2019), https://doi.org/10.1093/jue/juz002.

9. Celermajer et al., "Multispecies Justice."

10. Rosi Braidotti, *The Posthuman* (Oxford, UK: Polity Press, 2013); Val Plumwood, *Feminism and the Mastery of Nature*, Opening Out: Feminism for Today (London: Routledge, 2002), https://doi.org/10.4324/9780203006757.

11. David Schlosberg, "Justice, Ecological Integrity, and Climate Change," in *Ethical Adaptation to Climate Change*, ed. Allen Thompson and Jeremy Bendik-Keymer (Cambridge, MA: MIT Press, 2012), 165–184; David Schlosberg, "Ecological Justice for the Anthropocene," in *Political Animals and Animal Politics*, ed. M. Wissenburg and D. Schlosberg (London: Palgrave Macmillan UK, 2014), 75–89.

12. Danielle Celermajer, interview by Marina Alberti (MUltispecies Justice, April 16, 2025); Danielle Celermajer et al., *Institutionalising Multispecies Justice* (Cambridge, UK: Cambridge University Press, 2025).

13. Martha C. Nussbaum, *Frontiers of Justice*, The Tanner Lectures on Human Values (London: Belknap Press, 2006).

14. Danielle Celermajer et al., "Justice Through a Multispecies Lens," *Contemporary Political Theory* 19, no. 3 (September 2020): 475–512.

15. Houston et al., "Make Kin, Not Cities!"

Chapter 9

1. Italo Calvino, *Invisible Cities*, trans. W. Weaver (London: Picador, 1979).

2. Calvino, *Invisible Cities*, 32.

3. "Kirschner-Gerhart-1998-Evolvability."

4. "Kirschner-Gerhart-1998-Evolvability"; Joshua L. Payne and Andreas Wagner, "The Causes of Evolvability and Their Evolution," *Nature Reviews Genetics* 20, no. 1 (January 2019): 24–38; Richard Dawkins, "The Evolution of Evolvability," in *On Growth, Form and Computers*, ed. Thomas F. Hansen, David Houle, Mihaela Pavličev, and Christophe Pélabon (London: Elsevier, 2003), 239–255.

5. Mary Jane West-Eberhard, "Developmental Plasticity and the Origin of Species Differences," *Proceedings of the National Academy of Sciences of the United States of America* 102, suppl. 1 (May 3, 2005): 6543–6549.

6. James P. J. Hall, Michael A. Brockhurst, and Ellie Harrison, "Sampling the Mobile Gene Pool: Innovation via Horizontal Gene Transfer in Bacteria," *Philosophical Transactions of the Royal Society of London, Series B, Biological Sciences* 372, no. 1735 (December 5, 2017), https://doi.org/10.1098/rstb.2016.0424.

7. Peter R. Grant, *Ecology and Evolution of Darwin's Finches (Princeton Science Library Edition)*, Princeton Science Library 23 (Princeton, NJ: Princeton University Press, 1999).

8. Bateson, "Adaptability and Evolution."

9. Patrice Pottier et al., "Developmental Plasticity in Thermal Tolerance: Ontogenetic Variation, Persistence, and Future Directions," *Ecology Letters* 25, no. 10 (October 2022): 2245–2268.

10. Emilie C. Snell-Rood et al., "Mechanisms of Plastic Rescue in Novel Environments," *Annual Review of Ecology, Evolution, and Systematics* 49, no. 1 (November 2, 2018): 331–354; J. Hollander, E. Snell-Rood, and S. Foster, "New Frontiers in Phenotypic Plasticity and Evolution," *Heredity* 115, no. 4 (October 2015): 273–275.

11. Chris J. Blackman et al., "Genetic Adaptation and Phenotypic Plasticity Contribute to Greater Leaf Hydraulic Tolerance in Response to Drought in Warmer Climates," *Tree Physiology* 37, no. 5 (May 2017): 583–592.

12. Crawford Drury, "Resilience in Reef-Building Corals: The Ecological and Evolutionary Importance of the Host Response to Thermal Stress," *Molecular Ecology* 29, no. 3 (February 2020): 448–465.

13. Madhavi A. Colton et al., "Coral Conservation in a Warming World Must Harness Evolutionary Adaptation," *Nature Ecology & Evolution* 6, no. 10 (October 2022): 1405–1407.

14. "Singapore One Million Trees," 2020.

15. "Melbourne Urban Forest Strategy," 2024.

16. "Rotterdam Climate Change Adaptation Strategy" (City of Rotterdam, 2013).

17. "Copenhagen Flood Management Strategy," 2011.

18. "Amsterdam Circular Economy," accessed December 10, 2024, https://www.amsterdam.nl/en/policy/sustainability/circular-economy/.

19. "Paris 2024–2030 Climate Action Plan," accessed December 10, 2024, https://cdn.paris.fr/paris/2024/05/13/planclimat_synthese_en_web-qG4w.pdf.

20. "Paris Urban Nature Plans," accessed December 10, 2024, https://urbannatureplans.eu/pilot-cities/paris-france.

21. "Green City Freiburg," accessed December 10, 2024, https://greencity.freiburg.de/pb/1450158.html#:~:text=The%20City%20of%20Freiburg%20has,who%20live%20in%20the%20district.

22. "Lower Manhattan Coastal Resilience Project," accessed December 10, 2024, https://www.nyc.gov/site/lmcr/index.page.

23. David J. Earl and Michael W. Deem, "Evolvability Is a Selectable Trait," *Proceedings of the National Academy of Sciences of the United States of America* 101, no. 32 (August 10, 2004): 11531–11536.

24. Michael Barnett, Lena Zeller, and Paul B. Rainey, "Experimental Evolution of Evolvability," *BioRxiv* (May 3, 2024), https://doi.org/10.1101/2024.05.01.592015.

25. Christopher J. Graves et al., "Natural Selection Promotes Antigenic Evolvability," *PLoS Pathogens* 9, no. 11 (November 14, 2013): e1003766.

26. Earl and Deem, "Evolvability Is a Selectable Trait."

27. N. Colegrave and S. Collins, "Experimental Evolution: Experimental Evolution and Evolvability," *Heredity* 100, no. 5 (May 23, 2008): 464–470.

28. UN-Water, "SDG 6 Country Acceleration Case Study: Singapore," n.d.

29. Lai et al., "Volitional Activation of Remote Place Representations with a Hippocampal Brain-Machine Interface."

30. François Huber, *Nouvelles observations sur les abeilles, par François Huber, seconde édition revue, corrigée et considérablement augmentée, tome premier[-second]*—Primary Source Edition (Charleston, SC: Nabu Press, 2013).

31. Perry and Chittka, "How Foresight Might Support the Behavioral Flexibility of Arthropods."

32. Siddharth Chaudhary et al., "Bioinspired Technology in Society: Ethical and Architectural Innovations for Sustainable Development," *Technology in Society* 78, no. 102688 (September 2024): 102688.

33. Sheikh, Mitchell, and Foth, "Reparative Futures of Smart Urban Governance."

34. Katharina Sophia Mair, Johanna Irrgeher, and Daniela Haluza, "Elucidating the Role of Honey Bees as Biomonitors in Environmental Health Research," *Insects* 14, no. 11 (November 14, 2023): 874.

35. Helmut Schmitz and Herbert Bousack, "Modelling a Historic Oil-Tank Fire Allows an Estimation of the Sensitivity of the Infrared Receptors in Pyrophilous Melanophila Beetles," *PloS One* 7, no. 5 (May 21, 2012): e37627.

36. Vera Vasas et al., "Recording Animal-View Videos of the Natural World Using a Novel Camera System and Software Package," *PLoS Biology* 22, no. 1 (January 2024): e3002444.

37. Julien Dupeyroux, Julien R. Serres, and Stéphane Viollet, "AntBot: A Six-Legged Walking Robot Able to Home Like Desert Ants in Outdoor Environments," *Science Robotics* 4, no. 27 (February 13, 2019): eaau0307.

38. Marcel A. K. Jansen et al., "A Meta-Analysis of the Interactive Effects of UV and Drought on Plants," *Plant, Cell & Environment* 45, no. 1 (January 2022): 41–54.

39. Mair, Irrgeher, and Haluza, "Elucidating the Role of Honey Bees as Biomonitors in Environmental Health Research."

40. Zetian Zhao et al., "A Comprehensive Review on the Evolution of Bio-Inspired Sensors from Aquatic Creatures," *Cell Reports: Physical Science* 5, no. 7 (July 2024): 102064.

41. Florian Packmor et al., "Migratory Birds Can Extract Positional Information from Magnetic Inclination and Magnetic Declination Alone," *Proceedings: Biological Sciences* 291, no. 2034 (November 2024): rspb20241363.

42. Beth Mortimer et al., "Noise Matters: Elephants Show Risk-Avoidance Behaviour in Response to Human-Generated Seismic Cues," *Proceedings: Biological Sciences* 288, no. 1953 (June 30, 2021): 20210774.

Chapter 10

1. Lynn Margulis, *Microcosmos* (New York: Summit Books, 1986).

2. Zaal Kikvidze and Ragan M. Callaway, "Ecological Facilitation May Drive Major Evolutionary Transitions," *Bioscience* 59, no. 5 (May 1, 2009): 399–404.

3. David Sloan Wilson, "A Theory of Group Selection," *Proceedings of the National Academy of Sciences of the United States of America* 72, no. 1 (January 1975): 143–146; Vero Copner Wynne-Edwards, *Evolution Through Group Selection* (Philadelphia: Blackwell Science, 1986).

4. David Sloan Wilson and Edward O. Wilson, "Rethinking the Theoretical Foundation of Sociobiology," *The Quarterly Review of Biology* 82, no. 4 (December 2007): 327–348; Martin A. Nowak, Corina E. Tarnita, and Edward O. Wilson, "The Evolution of Eusociality," *Nature* 466, no. 7310 (August 26, 2010): 1057–1062.

5. William D. Hamilton, "The Genetical Evolution of Social Behaviour. I," *Journal of Theoretical Biology* 7, no. 1 (July 1964): 1–16; J. Maynard Smith, "Group Selection and Kin Selection," *Nature* 201, no. 4924 (March 1964): 1145–1147.

6. Peter Richerson et al., "Cultural Group Selection Plays an Essential Role in Explaining Human Cooperation: A Sketch of the Evidence," *The Behavioral and Brain Sciences* 39, no. e30 (January 2016): e30.

7. Edward O. Wilson, *The Social Conquest of Earth* (New York: W. W. Norton, 2012).

8. Wilson and Wilson, "Rethinking the Theoretical Foundation of Sociobiology."

9. Gregory J. Velicer and Yuen-Tsu N. Yu, "Evolution of Novel Cooperative Swarming in the Bacterium Myxococcus Xanthus," *Nature* 425, no. 6953 (September 4, 2003): 75–78; Paul B. Rainey and Katrina Rainey, "Evolution of Cooperation and Conflict in Experimental Bacterial Populations," *Nature* 425, no. 6953 (September 4, 2003): 72–74.

10. Charles J. Goodnight and Lori Stevens, "Experimental Studies of Group Selection: What Do They Tell Us About Group Selection in Nature?" *The American Naturalist* 150, suppl. 1 (July 1997): S59–S79.

11. William Swenson, David Sloan Wilson, and Roberta Elias, "Artificial Ecosystem Selection," *Proceedings of the National Academy of Sciences of the United States of America* 97, no. 16 (August 1, 2000): 9110–9114.

12. Charles J. Goodnight, "Heritability at the Ecosystem Level," *Proceedings of the National Academy of Sciences of the United States of America* 9, no. 16 (August 15, 2000): 9110–9114; Charles J. Goodnight, "Multilevel Selection: The Evolution of Cooperation in Non-kin Groups," *Population Ecology* 47, no. 1 (April 2005): 3–12.

13. Nowak, Tarnita, and Wilson, "The Evolution of Eusociality."

14. Edward O. Wilson and Bert Hölldobler, "Eusociality: Origin and Consequences," *Proceedings of the National Academy of Sciences of the United States of America* 102, no. 38 (September 20, 2005): 13367–13371; Wilson and Wilson, "Rethinking the Theoretical Foundation of Sociobiology."

15. Wilson and Wilson, "Rethinking the Theoretical Foundation of Sociobiology."

16. Wilson and Wilson, "Rethinking the Theoretical Foundation of Sociobiology."

17. Claire N. Spottiswoode, Keith S. Begg, and Colleen M. Begg, "Reciprocal Signaling in Honeyguide-Human Mutualism," *Science (New York, N.Y.)* 353, no. 6297 (July 22, 2016): 387–389.

18. Jessica E. M. van der Wal et al., "Safeguarding Human-Wildlife Cooperation," *Conservation Letters* 15, no. 4 (July 2022): e12886; Cram et al., "The Ecology and Evolution of Human-Wildlife Cooperation."

19. Cram et al., "The Ecology and Evolution of Human-Wildlife Cooperation."

20. Hussein Adan Isack, "The Biology of the Greater Honeyguide *Indicator indicator:* With Emphasis on the Guiding Behaviour" (PhD diss., University of Oxford, 1987); Husein Adrian Isack and H.-U. Reyer, "Honeyguides and Honey Gatherers: Interspecific Communication in a Symbiotic Relationship," *Science* 243, no. 4896 (1989): 1343–46.

21. Jessica E. M. van der Wal et al., "Cooperative Human Signals to Honeyguides Form Local Dialects," *People and Nature (Hoboken, N.J.)* (January 21, 2026), https://doi.org/10.1002/pan3.70234.

22. Claire N. Spottiswoode and Brian M. Wood, "Culturally Determined Interspecies Communication Between Humans and Honeyguides," *Science (New York, N.Y.)* 382, no. 6675 (December 8, 2023): 1155–1158.

23. Lynn Margulis, *Origin of Eukaryotic Cells: Evidence and Research Implications for a Theory of the Origin and Evolution of Microbial, Plant, and Animal Cells on the Precambrian Earth* (New Haven, CT: Yale University Press, 1970).

24. Eors Szathmary and John Maynard Smith, *The Major Transitions of Evolution* (Oxford, UK: WH Freeman Spektrum, 1995).

25. Eörs Szathmáry and John Maynard Smith, "The Major Evolutionary Transitions," *Nature* 374, no. 6519 (March 16, 1995): 227–232; Eörs Szathmáry, "Toward Major Evolutionary Transitions Theory 2.0," *Proceedings of the National Academy of Sciences of the United States of America* 112, no. 33 (August 18, 2015): 10104–10111.

26. Szathmáry and Smith, "The Major Evolutionary Transitions."

27. Wilson and Hölldobler, "Eusociality: Origin and Consequences"; Wilson and Wilson, "Rethinking the Theoretical Foundation of Sociobiology."

28. David Sloan Wilson et al., "Multilevel Cultural Evolution: From New Theory to Practical Applications," *Proceedings of the National Academy of Sciences of the United States of America* 120, no. 16 (April 18, 2023): e2218222120.

29. David Sloan Wilson, Elinor Ostrom, and Michael E. Cox, "Generalizing the Core Design Principles for the Efficacy of Groups," *Journal of Economic Behavior & Organization* 90 (June 1, 2013): S21–S32.

30. Wilson, Ostrom, and Cox "Generalizing the Core Design Principles for the Efficacy of Groups."

Chapter 11

1. Ebenezer Howard, *Garden Cities of Tomorrow* (London: Swan Sonnenschein & Company, 1902).

2. Sir Patrick Geddes, *Cities in Evolution: An Introduction to the Town Planning Movement and to the Study of Civics* (London: Williams & Norgate, 1915).

3. Ian L. McHarg, *Design with Nature*, The American Museum of Natural History (New York: Natural History Press, 1969).

4. Kevin Lynch, *The Image of the City* (Cambridge, MA: MIT Press, 1960); Kevin Lynch, *A Theory of Good City Form* (Cambridge, MA: MIT Press, 1981).

5. Beatley, *Biophilic Cities: Integrating Nature into Urban Design and Planning*; Beatley, *Handbook of Biophilic City Planning & Design*.

6. Edward O. Wilson, *Biophilia* (Cambridge, MA: Harvard University Press, 1984).

7. Stephen R. Kellert and Edward O. Wilson, ed., *The Biophilia Hypothesis*, 2nd ed. (Washington, DC: Island Press, 1993).

8. Timothy Beatley, *Biophilic Cities: Integrating Nature Into Urban Design and Planning* (Washington, DC: Island Press, 2011).

9. Bill Reed, "Shifting from 'Sustainability' to Regeneration," *Building Research and Information* 35, no. 6 (November 1, 2007): 674–680; Pamela Mang and Bill Reed, "Designing from Place: A Regenerative Framework and Methodology," *Building Research and Information* 40, no. 1 (January 1, 2012): 23–38.

10. Cristina Hernandez-Santin et al., "A Review of Existing Ecological Design Frameworks Enabling Biodiversity Inclusive Design," *Urban Science* 6, no. 4 (December 16, 2022): 95.

11. Chiara Catalano et al., "Smart Sustainable Cities of the New Millennium: Towards Design for Nature," *Circular Economy and Sustainability* 1, no. 3 (November 1, 2021): 1053–1086.

12. Wolfgang W. Weisser et al., "Creating Ecologically Sound Buildings by Integrating Ecology, Architecture and Computational Design," *People and Nature (Hoboken, N.J.)* 5, no. 1 (February 2023): 4–20.

13. Bruno Latour, *Reassembling the Social: An Introduction to Actor–Network-Theory* (Oxford: Oxford University Press, 2005); Donna Haraway, *Staying with the Trouble: Making Kin in the Chthulucene* (Durham, NC: Duke University Press, 2016); David Abram, *The Spell of the Sensuous: Perception and Language in a More-Than-Human World* (New York: Vintage Books, 1996); Sarah Whatmore, *Hybrid Geographies: Natures, Cultures, Spaces* (London: SAGE Publications, 2002); Anna Tsing, *The Mushroom at the End of the World: On the Possibility of Life in Capitalist Ruins* (Princeton, NJ: Princeton University Press, 2015).

14. Houston et al., "Make Kin, Not Cities!"

15. Christopher Alexander, "'A City Is Not a Tree,'" in *The Urban Design Reader* (London: Routledge, 2013), 152–166; Jane Jacobs, "The Death and Life of Great American Cities," in *Readings in Planning Theory* (Chichester, UK: John Wiley, Ltd, 2015), 94–109.

Conclusion

1. Laura Pereira et al., "Wakanda Phambili! African Science Fiction for Reimagining the Anthropocene Laura Pereira et al., "Wakanda Phambili! African Science Fiction for Reimagining the Anthropocene," in *Futures*, ed. Sandra Kemp and Jenny Andersson (Oxford: Oxford University Press, 2021).

2. Naomi Terry et al., "Inviting a Decolonial Praxis for Future Imaginaries of Nature: Introducing the Entangled Time Tree," *Environmental Science & Policy* 151 (January 1, 2024): 103615.

3. Terry et al., "Inviting a Decolonial Praxis for Future Imaginaries of Nature."

4. Henrik Ernstson and Sverker Sörlin, *Grounding Urban Natures: Histories and Futures of Urban Ecologies* (Cambridge, MA: MIT Press, 2019).

Index

Note: Figures and tables are indicated by italicized page numbers.

Publisher contact:
The MIT Press
Massachusetts Institute of Technology
77 Massachusetts Avenue, Cambridge, MA 02139
mitpress.mit.edu

EU Authorised Representative:
Easy Access System Europe, Mustamäe tee 50,
10621 Tallinn, Estonia
gpsr.requests@easproject.com

Printed by Integrated Books International,
United States of America